VITAL SIGNS 2000

VITAL SIGNS 2000

The Environmental Trends That Are Shaping Our Future

Lester R. Brown

Michael Renner

Brian Halweil

Editor: Linda Starke

with
Janet N. Abramovitz
Seth Dunn
Christopher Flavin
Hilary F. French
Gary Gardner
Nicholas Lenssen
Lisa Mastny
Ashley T. Mattoon
Anne Platt McGinn
Sarah Porter
Sandra Postel
David M. Roodman
Payal Sampat
Michael Scholand
Molly O. Sheehan

W.W. Norton & Company
New York   London

The text of this book is composed in Garth Graphic
with the display set in Industria Alternate.

Composition by the Worldwatch Institute; manufacturing by the Haddon Craftsmen, Inc.
Book design by Charlotte Staub.

ISBN 0-393-32022-7 (pbk)

W.W. Norton & Company, Inc.
500 Fifth Avenue, New York, NY 10110
W.W. Norton & Company Ltd.
10 Coptic Street, London WC1A 1PU

1234567890

Worldwatch Database Disk Subscription

Worldwatch offers the data from all graphs and tables contained in this book, as well as graphs and tables in all other Worldwatch publications of the past two years, on 3½-inch disks for use with PC or Macintosh computers. This includes data from the State of the World *and* Vital Signs *series of books, other Norton/Worldwatch books, Worldwatch Papers, and* WORLD WATCH *magazine. Subscribers can import these data into spreadsheets, presentation software, or word processing programs. Database Disk subscribers receive the disk plus a copy of* Vital Signs 2000, *as well as a copy of* State of the World 2001 *when it is published in January. The data (in spreadsheet format) are provided as Microsoft Excel 5.0/95 workbook (*.xls) files. Users must have spreadsheet software that can read Excel workbooks for Windows. The Worldwatch Database Disk Subscription costs just $89 (plus $4 shipping and handling). Phone: (800) 555-2028. Fax: (202) 296-7365. E-mail: <wwpub @worldwatch.org>. Mastercard, Visa, or American Express accepted.*

Visit our Web site at www.worldwatch.org

CONTENTS

Part One: KEY INDICATORS

Part Two: SPECIAL FEATURES

ACKNOWLEDGMENTS

We are grateful to the W. Alton Jones Foundation for its generous support for *Vital Signs 2000*. Since we first published *Vital Signs* in 1992, the book has become one of the pillars of our work and a valued reference source worldwide.

Some of the data that we present in *Vital Signs* are the fruit of research for our other publications. At the same time, projects like *State of the World*, the Worldwatch Papers, and WORLD WATCH magazine stimulate ideas for new *Vital Signs* indicators and features. These projects are supported by additional foundations and individual donors. We thank the Compton Foundation, the Geraldine R. Dodge Foundation, the Ford Foundation, the William and Flora Hewlett Foundation, the John D. and Catherine T. MacArthur Foundation, the Charles Stewart Mott Foundation, the Curtis and Edith Munson Foundation, the David and Lucille Packard Foundation, the Rasmussen Foundation, Rockefeller Financial Services, the Summit Foundation, the Turner Foundation, the Wallace Genetic Foundation, the Wallace Global Fund, the Weeden Foundation, and the Winslow Foundation. In addition, we would like to acknowledge the support of the more than 1,800 individuals who provided financial support through the Friends of Worldwatch program last year. Our special appreciation goes to the members of our Council of Sponsors: Tom and Cathy Crain, Roger and Vicki Sant, Robert Wallace and Raisa Scriabine, and Eckart Wintzen.

More than any other Worldwatch publication, *Vital Signs* brings together the analytical and writing talents of our entire research staff. In addition to current staff, we are joined in this edition by several Worldwatch alumni. Worldwatch senior fellow Sandra Postel pitched in from Amherst, Massachusetts, where she directs the Global Water Policy Project. Nicholas Lenssen continues to contribute his invaluable annual update on nuclear power from Boulder, Colorado. And former interns Sarah Porter and Michael Scholand wrote on international debt, pesticide trade, and efficient lighting.

The consistency of style, tone, and appearance throughout the book is testament to the editing qualities and experience of veteran editor Linda Starke, who shepherded 48 manuscripts written by 18 authors from draft form to final published version. As in previous years, Art Director Liz Doherty was in charge of desktop production of *Vital Signs*. We are grateful for her cheerful and fast design work, even as other responsibilities compete intensely for her time. Librarian Lori Brown and Library Assistant Jonathan Guzman are instrumental in keeping authors well supplied with books, reports, and other research materials. Christine Stearn incorporates all the tables and figures in the print version of *Vital Signs* into the Worldwatch Database disk and manages our Web site, including the download page for individual *Vital Signs* pieces.

We are grateful to them as well as to all the other Worldwatch staffers whose behind-the-scenes work makes this book possible. They include Reah Janise Kauffman, who assists with fundraising and, with Mary

Redfern, is our liaison with domestic and foreign publishers; our operations team of Barbara Fallin, Suzanne Clift, and Sharon Lapier; our communications team of Richard Bell, Mary Caron, and Liz Hopper; and our publications sales team of Millicent Johnson and Joseph Gravely. Without their hard work, we could not publish and market *Vital Signs* or disseminate the information it contains.

All contributions to this book were reviewed by inhouse staff as well as by a number of outside experts. For particular help with data requests, advice, or feedback on drafts, we would like to thank Earle Amey, Neil Austriaco, Wasantha Bandarage, Nils Borg, Barry Bredenkamp, Colin Couchman, Victor Gaigbe-Togbe, Jenni Gainsborough, Bernward Geier, Catherine Greene, Tim Halliday, Paul Hunt, Frank Jamerson, Nic Lampkin, Don MacKay, Paul Maycock, Masa Nagai, Mika Ohbayashi, Thomas Rabehl, Jamie Reaser, José Santamarta, Vladimir Slivyak, Carrie Smith, Stefan Speck, Russell Sturm, Arnella Trent, Andreas Wagner, Neff Walker, Tom Wigley, Helga Willer, Angelika Wirtz, Bock Cheng Yeo, and John Young.

Once the edited manuscript leaves our hands, Amy Cherry, Nomi Victor, Andrew Marasia, and their colleagues at W.W. Norton & Company ensure that it is printed in record time. We sincerely thank them for their unwavering support.

Most of our waking hours are dedicated to building a sustainable, more just, and peaceful world—one that we can be proud to pass on to our children and grandchildren. During the past year, the extended Worldwatch family was enriched by three joyful additions. We welcome Benjamin Roodman, Ileanna Guzman, and Jack McGinn by dedicating this book to them and their generation.

Lester R. Brown
Michael Renner
Brian Halweil

FOREWORD

In this ninth edition of *Vital Signs*, we again put our finger to the world's pulse by compiling a wide-ranging collection of trends that identify both problems and progress in the quest for a sustainable society. Ten of this edition's 48 indicator and feature pieces cover new topics—issues not discussed in previous editions. They are pesticide trade, tourism, organic farming, groundwater pollution, ice melting, endocrine disrupters, remote sensing, wind energy employment, tuberculosis, and prison populations.

As in past editions of *Vital Signs*, this year's survey documents both alarming situations and encouraging developments. On the plus side, for instance, the growth in wind power, solar photovoltaics, and energy-efficient light bulbs is gathering enormous momentum. On the downside, consumption of virgin wood pulp and paper continues to rise, more than offsetting impressive gains in paper recycling and recovery.

But irrespective of whether the trend is positive or negative, the picture that emerges from the broad panoply of topics is one of astonishing disparity among the world's people—inequalities of wealth, power, opportunities, and survival prospects. Even as the world increasingly becomes "one" by dint of trade, investment, travel, and Internet connections, humanity continues to be plagued by deep divisions. These fault lines often run between the global North and South, but frequently are found also within nations and between men and women.

Ironically, even as disparity grows, diversity on the planet seems to diminish—whether it be in the realm of biological resources (the decline of amphibians and the rising monoculture of industrial agriculture), culture (the homogenization of the world via mass entertainment and the Internet), economics (globe-straddling conglomerates as a result of corporate mergers), or transportation (the dominance of a car-centered transport system).

Some argue that the Internet will help overcome many of our current predicaments. But will the new information and communication technologies assist in moving decisively toward sustainability, or will they perpetuate and supercharge the current mass consumption system? Will they allow leapfrogging by those left behind, or will they widen the gulf between haves and have-nots? At least for the time being, we see disparity as much in the emerging digital world as in the physical world. Although the situation is a fast-changing one, only 4 percent of the world's people are connected to the Internet. Some 87 percent of current Internet users live in industrial countries. Africa, by contrast, is largely outside the fold.

While shopping via the Web has been reduced to little more than the bit of eye-hand coordination required for a series of mouse-clicks, the impact of a consumption-heavy lifestyle on the environment has decidedly not been lessened. Whereas the digital age promises "everything, all the time, right away," the momentum of the industrial system and of growing human numbers is such that climate change, soil degradation, overpumping of groundwater, and deforestation cannot be easily reversed.

This is not to argue that the advances made in computers, satellites, and associated technologies cannot be of help in the struggle for sustainability. Indeed, as we point out in this edition of *Vital Signs*, information and communications technologies are improving our ability to monitor Earth's condition, enhancing our understanding of complex natural processes, and permitting those dedicated to saving the planet to be in touch with each other and to communicate their urgent message. But it is evident that in the quest for sustainability, we need to be mindful of the goals of solidarity and equity as well.

As we have just done with this brief look through the lens of disparity, we invite the reader to engage in cross-cutting comparisons of material in this book. Although the individual indicators and features are written as stand-alone pieces, they truly are part of a complex mosaic.

Some combinations are obvious. For instance, we augment our staple discussion of wind power trends with an analysis of the employment benefits of this renewable energy source. We have a discussion of world paper production and an analysis of paper recycling trends. We include both a general article on international trade and a piece on imports and exports of pesticides. And we discuss carbon emissions and rising global temperatures as well as the various impacts of climate change: storm damage, melting of Earth's ice cover, and the extinction of the Golden Toad and other amphibians that serve as a kind of barometer of Earth's health.

Similarly, the indicator on international trade can be read side by side with the feature on corporate mergers. Intra-firm trade—the flow of commodities, manufactured goods, and services from a subsidiary of a transnational corporation in one country to another subsidiary in a second country—accounts for about one third of world trade.

Readers interested in our regular reporting on food and agricultural resource trends should peruse the features on groundwater depletion and pollution. They may also want to consult entries on organic farming and

transgenic crop area, in many ways two diametrically opposite trends in food production. Although the area used to raise genetically modified crops is roughly six times that devoted to organic methods, the latter may ultimately prove the more significant development. Transgenic crops are almost exclusively grown in only three countries, and public resistance to them is rising around the globe. Organic farming methods, in contrast, are gaining favor in many countries, in part because they help reduce groundwater pollution.

Readers may further want to consult pieces on pesticide trade and endocrine disrupters. Pesticides such as atrazine, DDT, and endosulfan are among the chemicals that interfere with the human endocrine system, which regulates many of the body's vital processes.

A Part Two feature on remote sensing is pivotal, relating to several other vital signs in the book. Satellites are an important tool in the effort to better understand the dynamics of weather patterns (and help predict severe events leading to the massive storm damages discussed in Part One), monitor the complex global climate system (also discussed in Part One), and provide readings of changes in the polar ice regions (described in Part Two). The remote sensing piece is also of particular relevance to the information contained in the environmental treaties article, as such agreements often stand or fall on the issue of adequate monitoring.

The piece on tuberculosis also has relevance for several others. It is illuminating to read in conjunction with the one on AIDS. AIDS has been the single largest factor behind the surge in TB infections. Because the AIDS virus weakens the human immune system, an HIV-positive individual is at much greater risk to develop TB. The tremendous upswing in international tourism has contributed to the global spread of TB, as has the growth in the world's refugee population. Finally, the situation in the world's prisons has relevance for TB as well, as prisons are frequently breeding grounds for the disease: overcrowding combined with inadequate sanitation, nutrition, and health care facilitates the spread of TB, AIDS, and other diseases.

* * *

As in previous years, we provide all the data contained in the tables and figures of this book in an updated version of the Worldwatch Database Disk. Individual *Vital Signs* indicators can be downloaded in Adobe Acrobat Reader (.pdf) format from our Web site, at <www.worldwatch.org/titles/tvs. html>. Elsewhere on our Web site, we list the foreign language editions of *Vital Signs* at <www.worldwatch.org/foreign/index.html>.

On behalf of our coauthors, thank you for your interest in *Vital Signs 2000*. Please let us know by e-mail (<worldwatch@worldwatch. org>), fax (202-296-7365), or regular mail if you have any suggestions for improving future editions or for new indicators that we should consider.

Lester R. Brown
Michael Renner
Brian Halweil
March 2000

Worldwatch Institute
1776 Massachusetts Ave., N.W.
Washington DC 20036

VITAL
SIGNS
2000

OVERVIEW
The Acceleration of Change

Lester R. Brown

We have noted in earlier editions of *Vital Signs* that history appeared to be accelerating, that everything was moving faster. The last year of the old century was no exception. Records were being set on so many fronts that we could scarcely keep track. In 1999, world population passed 6 billion, adding the last billion in a record 12 years. And India's population reached 1 billion. Neither demographic milestone was a cause for celebration.

During the last half of the twentieth century, world population increased from 2.5 billion to 6 billion, with most of the increase coming in the developing world. In country after country, the population was outrunning the water supply. The demand for firewood and lumber was outrunning the sustainable yield of forests. And the demand for food was outrunning the cropland area.

The world ended the twentieth century on a strong economic note. The global economy had just completed a sixfold expansion in 50 years. Powering this was a fourfold growth in fossil fuel use, accompanied by a similar increase in carbon dioxide (CO_2) emissions. Each year since systematic air sampling began, atmospheric CO_2 levels have moved to a new high, climbing from 317 parts per million (ppm) in 1959 to 368 ppm in 1999.

This 16-percent rise in the concentration of CO_2, the principal greenhouse gas, was accompanied by a record rise in temperatures, which contributed to some of the most destructive storms and floods on record. And as Earth's temperature rises, its ice cover is melting. Scientists report that the Arctic sea ice has thinned by 40 percent over the last three decades. Ice sheets around the Antarctic Peninsula have broken up, yielding Delaware-sized icebergs. The vast snow-ice mass in the Himalayas—the third largest after that of the two poles—is melting rapidly.

Even as these signs of climate disruption were multiplying, signs of a new climate-benign energy economy based on renewable energy resources were emerging. While coal production dropped by 3 percent in 1999, wind electric generation increased by 39 percent as new wind farms came on line in Minnesota, Iowa, Texas, Wyoming, and Oregon in the United States, in Spain, in northwestern Europe, and in China. Solar cell production, including a large component of solar roofing materials, jumped by 30 percent in 1999.

These were encouraging signs that the world is beginning to respond to the environmental threats that promise to undermine our future, but the gap between what we need to be doing to reverse the environmental deterioration of the planet and what we are actually doing continues to widen. Many have come to expect that the progress in improving the human condition that marked the last half of the twentieth century would continue during the twenty-first, but in sub-Saharan Africa—where the capacity to respond to new threats has been weakened by continuing rapid population growth—progress is being reversed. The HIV epidemic in this region has reached epic proportions, threatening to take more lives during the first two decades of this century than World War II did in the last century.

ENERGY TRANSITION ACCELERATES

The transition from fossil fuels to a solar/hydrogen energy economy accelerated sharply in 1999. (See Table 1.) The burning of coal, the fossil fuel that launched the industrial era, declined by 3 percent in 1999; oil increased by 1 percent; and natural gas, the cleanest burning, least climate-disruptive of the three fossil fuels, expanded by 3 percent. (See pages 52–53.) Nuclear power, once seen as the energy source of the future, barely maintained its expansion in 1999 with a growth of 0.4 percent. (See pages 54–55.) Meanwhile, world wind generating capacity grew by 39 percent and sales of solar cells by 30 percent. (See pages 56–59.)

World coal consumption is the first of the fossil fuels to peak and begin to decline. After reaching a historic high in 1996, it has dropped by 6 percent and is expected to continue declining as the shift to natural gas and renewables gains momentum. Some estimates have oil production peaking before the end of this decade. Only natural gas, now viewed by many as the transition fuel from the fossil era to the solar/hydrogen era, is likely to continue growing for an extended period.

Coal consumption is declining sharply in the United Kingdom, where the Industrial Revolution began, and in China, the world's largest user of coal. Cuts in subsidies for coal in China and the closing of inefficient state-owned mines have both contributed to its declining use. These changes are being driven by air pollution in Chinese cities, which include some of the most polluted urban areas in the world. By shifting from coal to natural gas, cities can begin to reduce the urban air pollution that has claimed literally millions of lives in China in recent years.

As part of its long-term planning, China is building a new pipeline from the gas fields discovered in its northwest to Lanzhou in Gansu Province. China has also approved the import of natural gas and is now planning to build a pipeline linking Russia's Siberian gas fields with Beijing and Tianjin, two leading industrial cities.

The shift in the fortunes of nuclear power could hardly be more dramatic. In the 1980s, world nuclear generating capacity expanded by 140 percent; during the 1990s, it expanded by less than 5 percent. The energy source that was to be "too cheap to meter" is now too costly to use. Wherever electricity markets are opened to competition, nuclear power is in trouble. Its use is likely to peak within the next three years.

Nuclear power plant closings are now under way or slated in the years immediately ahead in many countries, including Bulgaria,

TABLE 1. TRENDS IN ENERGY USE, BY SOURCE, 1990–99[1]	
ENERGY SOURCE	ANNUAL RATE OF GROWTH (percent)
Wind power	+ 24.2
Solar photovoltaics	+ 17.3
Geothermal power[2]	+ 4.3
Natural gas	+ 1.9
Hydroelectric power[2]	+ 1.8
Oil	+ 0.8
Nuclear power	+ 0.5
Coal	− 0.5

[1]Trends measured in varying units: installed generating capacity (megawatts or gigawatts) for wind, geothermal, hydro, and nuclear power; million tons of oil equivalent for oil, natural gas, and coal; megawatts for shipments of solar photovoltaic cells.
[2]1990–98 only.
SOURCE: See pages 52–59.

Germany, Kazakhstan, the Netherlands, Russia, the Slovak Republic, Sweden, and the United States. In three countries once solidly committed to nuclear power—France, China, and Japan—nuclear power is losing its appeal. France has extended its moratorium on new nuclear power plants. China has said it will not approve any additional plants for the next three years. Japan's once ambitious nuclear program is in trouble. A serious accident in September 1999 at a nuclear fuel fabrication plant north of Tokyo has reinforced the fast-growing anti-nuclear movement in Japan.

Meanwhile, the use of wind and solar cells, the cornerstones of the new energy economy, is growing by leaps and bounds. One of the attractions of wind-generated electricity is its falling cost. With the new advanced design wind turbines, electricity is typically being generated at 4–6¢ per kilo-watt-hour, one fourth the cost of a decade ago and a figure that is competitive with traditional energy sources. Indeed, annual additions of wind capacity during the late 1990s exceeded those of nuclear power. In effect, the torch is passing to a new generation of energy technologies.

Germany has emerged as the world leader in wind electrical generating capacity, with the United States in second place. Major new wind farms have begun operation over the last two years in Minnesota, Iowa, Texas, Wyoming, and Oregon. This growth in wind electric generation in the Corn Belt and the Great Plains is providing farmers and ranchers with welcome supplemental income. Indeed, the Great Plains has enormous wind-generating potential, making it the Saudi Arabia of wind power.

Europe is moving quickly to develop its wind energy resources. Denmark, the world leader in advanced design wind turbine manufacturing, continues to add new capacity. The country where wind power is growing fastest is Spain. Starting from zero four years ago, Spain moved into second place in terms of new wind installations in 1999 with 750 megawatts, trailing only Germany at 1,570 megawatts. In early 2000, Energia Hidro-electrica de Navarra, the leader in wind energy development in Spain, announced an order for some 1,400 megawatts worth of wind turbines—the largest order ever placed.

European countries are now excited by the offshore potential for generating wind. A new study indicates that in the coastal regions of the North Sea and the Baltic Sea, out to a depth of 30 meters, there is enough harnessable wind to satisfy the continent's electricity needs.

In addition to being a climate-benign source of energy, wind power is also labor-intensive. In Germany, for example, where wind supplies 2 percent of electricity generation, in 1998 the industry employed an estimated 15,000 workers in the manufacture, installation, and operation of wind turbines. (See pages 146–47.) By contrast, nuclear power, which supplies 31 percent of electricity, offered only 40,000 jobs. In Europe, where double-digit unemployment rates are not uncommon, the large number of jobs created in a wind power energy economy are a definite plus.

The growth in solar cell manufacturing is also accelerating, jumping from an average of 16 percent a year from 1990 to 1998 to 30 percent in 1999. Japan, the United States, and several countries in Europe now have solar cell manufacturing facilities. The largest producer in the world today is BP Solarex. In Germany, Royal Dutch Shell opened a 25-megawatt, fully automated production facility. The big advance in solar cell potential came with the development by the Japanese of a solar roofing material, which means that the roof can become the power plant for the building.

As the world turns to new sources of energy, new technologies are substantially boosting the efficiency of energy use. Among the more dramatic of these are compact fluorescent lamps, light bulbs that provide the same amount of lighting as an incandescent bulb but use only one fourth as much electricity. The estimated 1.3 billion compact fluorescents in use today are operating on 20,000 megawatts of electricity, a huge saving over the 80,000 megawatts of capacity that would be needed to light the same number of incandescent bulbs. (See pages 60–61.)

CLIMATE CHANGE BUILDING MOMENTUM

Earth's average temperature in 1999 was down somewhat from 1998, which was the highest in the last century. (See pages 64–65.) Nonetheless, 1999 was the seventh warmest year since 1866, when continuous recordkeeping began. (See Figure 1.) The average temperature in 1998 was well above trend because of the El Niño warming of Pacific equatorial waters, and conversely it dropped below trend in 1999 due to La Niña, the flip side of the El Niño effect. As atmospheric CO_2 levels rise, Earth's average temperature is also rising.

Carbon emissions from fossil fuel burning have been more or less flat for the last three years at roughly 6.3 billion tons per year. (See pages 66–67.) At this level, however, they far exceed nature's capacity to fix carbon, thus pushing atmospheric concentrations of CO_2 higher. This rise in CO_2 levels, which have climbed higher every year since air sampling began in 1959, has become one of the most predictable of all the trends shaping our future.

Some of the expected effects of climate change, such as more destructive storms and the melting of Earth's ice cover, are now becoming evident. Weather-related damage in 1999 totaled $67 billion worldwide, the second highest after the 1998 figure of $93 billion. (See pages 76–77.) Weather-related damage worldwide during the 1990s was more than five times the figure during the 1980s.

Among the more devastating storms in 1999 was one in Venezuela that claimed 30,000 lives and destroyed an estimated $15 billion worth of property. One of Latin America's worst natural disasters in this century, it was the product of not only an uncommonly destructive storm but also extensive deforestation and construction in high-risk areas. A series of wind storms that hit Western Europe, importantly France, Germany, Spain, and Switzerland, did $9.6 billion worth of damage in late 1999. And a super cyclone with winds of 300 kilometers (190 miles) per hour that moved out of the Bay of Bengal into the East Indian state of Orissa in October took 15,000 lives.

Another consequence of higher temperatures is the melting of ice, a process that accelerated during the 1990s. Arctic sea ice, for example, has thinned by a staggering 40 percent within the last 30 years. (See pages 126–27.) The Antarctic's continent-sized ice sheet, which is on average 2.3 kilometers thick, is relatively stable, but the ice shelves—the part that floats on the surrounding seas—are melting rapidly. Three ice shelves along the West Antarctic peninsula—the Wordie, the Larsen A, and the Prince Gustav—have broken up entirely. Delaware-sized icebergs that have broken off are threatening ships in the area.

Ice is also melting rapidly in subpolar regions and mountains. For example, the Alps have lost 50 percent of their glacial mass over the last century. In the United States, Glacier

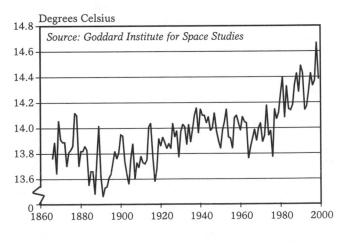

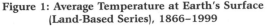

Figure 1: Average Temperature at Earth's Surface (Land-Based Series), 1866–1999

National Park has lost two thirds of its glaciers since 1850. Those remaining could disappear within the next 30 years. The Himalayan snow-ice mass is also melting. Feeding all the major rivers of Asia, including the Ganges, Indus, Mekong, Yangtze, and Yellow rivers, it is projected to shrink by 20 percent over the next 35 years.

FOOD TRENDS MIXED

World grain production in 1999 fell by 1 percent from the year before, dropping the per capita supply by more than 2 percent. (See pages 34–35.) Among the big three grains—wheat, corn, and rice—production of wheat and corn dropped by 1 percent each while rice increased by 1 percent.

World grain production per person dropped by more than 2 percent in 1999. This drop extended a decline that has been under way since 1984, one that has reduced per capita grain production worldwide by some 10 percent.

Trends contrast widely among regions. Much of the per capita decline has come in the republics of the former Soviet Union, including the two large ones—Russia and the Ukraine. The other major region suffering a decline is Africa. Continuing rapid population growth, steadily shrinking cropland area per person, and the loss of soil from erosion have all contributed to the region's deteriorating food situation.

One manifestation of growing demand for animal protein (see Figure 2) has been the extraordinary ninefold growth in the world soybean harvest from 1950 to 1999, a jump from 17 million to 154 million tons. (See pages 36–37.) A modest amount of soybean meal added to grain consumed by livestock and poultry greatly enhances the efficiency of the grain used. This ninefold expansion contrasts with a threefold growth in the world grain harvest during the same period.

Although the soybean originated in China,

it has found an ecological and economic niche in the United States, which today produces nearly half of the world's soybeans. Indeed, in 1999, the U.S. soybean harvested area eclipsed that of corn and wheat, traditionally the two leading crops, for the first time in history.

World meat production, which increased by 1 percent in 1999, has now risen for 41 consecutive years—making it one of the most predictable of the world's food consumption trends. (See pages 38–39.) Of the three meats that dominate human diets, beef and pork each increased by 0.5 percent in 1999 while poultry increased by nearly 3 percent, accounting for most of the growth in the world's meat supply.

Twenty years ago, the United States was

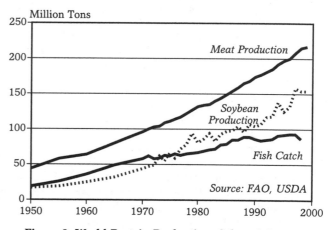

Figure 2: World Protein Production, Selected Sources, 1950–99

the world's leading consumer of meat. But this has now changed. After the economic reforms in China in 1978, Chinese incomes multiplied fourfold within two decades. As a result, in 1999 China consumed 55 million tons of meat compared with 34 million tons in the United States.

Two of the resource systems that support traditional sources of animal protein in the human diet—rangelands, which account for much of the world's beef and mutton production, and oceanic fisheries, which supported

21

a fivefold growth in the world fish catch between 1950 and 1998—are approaching their productive limits.

One reason world beef production increased little during the 1990s is that most of the world's rangelands are being grazed at or beyond capacity. A similar situation exists in fisheries—some 60 percent of all oceanic fisheries are now being fished at or beyond capacity. In 1998, the last year for which data are available, the catch dropped from 93 million tons to 86 million tons. (See pages 40–41.) This precipitous drop of nearly 8 percent reflected a decline in catch in some overworked fisheries and the El Niño weather event, which warmed the eastern Pacific waters, sharply reducing fish stocks there.

While the oceanic catch was dropping, fish farming continued to expand, going from 29 million tons in 1997 to 31 million tons in 1998. Two thirds of world output is concentrated in China, where people rely on several species of carp for much of their fish consumption.

UNSUSTAINABLE OUTPUT
GROWING

A small but growing share of the world grain harvest is being produced with the unsustainable use of land and water. In 1999, world grain area totaled 674 million hectares, down some 8 percent from the historical high of 732 million hectares in 1981. (See pages 44–45.) Part of this decline was because highly erodible cropland was being returned to grass or trees. This is most evident in the United States, where some 12 million hectares—nearly one tenth of U.S. cropland—was returned to grass or trees under the Conservation Reserve Program. In China, the leading loss of cropland is also the conversion of highly erodible cropland to its original vegetative cover.

Some countries are losing cropland to nonfarm uses. Between 1982 and 1997, more than 5 million hectares of U.S cropland were converted to other uses. An estimated one fifth of China's cropland loss is attributed to the construction of roads, factories, and homes.

The other major source of unsustainable food production is that resulting from the overpumping of underground aquifers. Overpumping worldwide is now conservatively estimated at 160 billion cubic meters of water per year. (See pages 122–23.) Using the rule of thumb to convert water into grain of a thousand cubic meters of water to produce a ton of grain, this would total 160 million tons of grain. Stated otherwise, if we were to decide this year to stabilize water tables throughout the world, the world grain harvest would drop by something like 160 million tons. At average world consumption of roughly one third of a ton of grain per person per year, this would feed 480 million people. In effect, 480 million of the world's current population of 6 billion are being fed with food produced with the unsustainable use of water.

In summary, the sustainability of world food production and of the population that depends on it is being threatened by the loss of cropland from erosion and conversion to nonfarm uses and by the overpumping of aquifers.

THE PRODUCTIVITY CHALLENGE

One of the keys to the tripling of the world grain harvest over the last half-century was the rise in land productivity. Farmers in a growing number of countries, however, are finding it difficult to sustain this historically rapid growth. Among them are rice farmers in Japan and wheat farmers in the United States and Mexico. In part this is because of the declining response of crops to additional applications of fertilizer. Many high-yielding crops are simply approaching their physiological capacity to absorb additional nutrients.

Efforts to maintain land productivity are further complicated by the urbanization of world population, which has led to a wholesale disruption of nutrient recycling. One reason world fertilizer use increased from 14 million tons in 1950 to 134 million tons in 1999 (see pages 46–47), nearly a 10-fold increase, was as a replacement for the nutrients being lost from farmland as crops are exported to cities where the nutrients enter local sewage systems, often ending up in a nearby river or the ocean.

The United States, for example, exports roughly 100 million tons of grain per year and with it all the nutrients in the grain. Without fertilizer to replace these nutrients, land productivity in major grain-producing states like Kansas and Iowa would be gradually declining over time as a result of nutrient depletion.

One hope for raising land productivity that is widely heralded by the seed-producing industry, namely the genetic modification of crops by transferring germplasm from other species, has not materialized. Thus far the growth in the area planted to genetically modified crops, which expanded from scratch in 1995 to 40 million hectares in 1999, has had no measurable effect on crop production. (See pages 118–19.) The use of these genetically modified crops has, however, affected pesticide use. It has reduced insecticide use on cotton and, to a lesser degree, on corn, and has dramatically raised herbicide use on soybeans.

Even as the economic gains for farmers from these genetically modified crops are perhaps less than expected, there is mounting concern among consumers and environmentalists about the effects on human health and on the environment. As a result of this growing concern, the U.S. area planted to genetically modified crops in 2000 is likely to drop by some 15–25 percent.

Another agricultural trend, the shift to organic farming, is continuing to gain momentum, reaching an estimated 7 million hectares in 1999. (See pages 120–21.) Aside from eliminating the risk of pesticide contamination of food, organic farming also reduces pesticide and nutrient runoff from cropland.

The area of land that is farmed organically, which is less than 1 percent of world cropland, contrasts sharply with the growth in the area planted to genetically modified crops, which reached 40 million hectares in just four years. In 2000, the area farmed organically is projected to continue expanding while that planted to genetically modified crops is expected to shrink. Ironically, neither trend

appears to be contributing to any growth in the world food supply.

ECONOMIC TRENDS MIXED

In 1999, the world economy expanded by 3 percent, up from 2.5 percent the year before. (See pages 70–71.) The $40.5 trillion worth of goods and services produced in 1999 was up more than sixfold from the $6.3 trillion output of goods and services in 1950.

The global economy is becoming huge compared with the capacities of Earth's ecosystems to supply basic goods, such as forest products, fresh water, and seafood. The $1.2 trillion expansion in output during 1999 exceeded the growth in the global economy during the entire nineteenth century.

While the global economy was expanding in 1999, international trade was virtually unchanged, thus slightly reducing the share of world economic output traded in 1999. (See pages 74–75.) According to this key indicator, globalization declined slightly in 1999.

World trade consists of both goods and services. The principal services include tourism, banking, insurance, and licenses for intellectual property, such as software and movies. While international trade in goods barely increased in 1999, international tourism rose 3 percent. (See pages 82–83.) Although this is below the average annual rate of growth of 7 percent since 1950, it brought the number of international tourist arrivals in 1999 to 657 million. As tourism, which today accounts for 12 percent of global economic activity, has expanded since 1980, the number of hotel beds worldwide has jumped by more than 80 percent and now exceeds 29 million. Each hotel room added typically creates at least one new job.

Although earnings in developing countries from international tourism have been rising rapidly, they have not been sufficient to avoid a rise in the external debt of developing countries. Expanding 5 percent in 1998, this debt grew faster than both the world economy and international trade. (See pages 72–73.) For some of the most heavily indebted poor countries, servicing external debt is siphoning resources away from meeting basic needs. In Zambia, for example, 30 percent of government spending is used to pay off foreign debt while only 10 percent is available to invest in health, education, and other basic social services. To help alleviate this financial stress, industrial-country governments have agreed to write off roughly two thirds of the official debt owed by the poorest countries.

In the transportation sector, global passenger car production expanded 3 percent in 1999, reaching an all-time high of 39 million vehicles. (See pages 86–87.) In North America, the share of automobile sales accounted for by light trucks, sport utility vehicles, and pickup trucks increased from 20 percent in 1975 to 46 percent in 1999. One consequence of this is a decline in fuel efficiency of the U.S. passenger vehicle fleet from 25.9 miles per gallon in the early 1980s to 23.8 miles per gallon in 1999.

While automobile production was reaching an all-time high, bicycle production sagged for the third year in a row. (See pages 88–89.) In 1995, when output peaked at 107 million, three times as many bikes were produced as cars. With bicycle manufacturing dropping to 79 million in 1998, the margin of bicycles over cars has been reduced from three-to-one to two-to-one. (See Figure 3.)

The principal reason for the decline in bicycle manufacturing has been the saturation of the huge bicycle market in China that occurred as economic reforms and rising affluence enabled literally hundreds of millions of Chinese to buy bicycles during the 1980s and early 1990s. Once this market demand was met, bicycle manufacturing and production dropped sharply in China.

Elsewhere, however, many cities are turning to bicycles partly because of frustration with automobile traffic congestion and pollution. Bogota, for example, is investing heavily in bicycle infrastructure to encourage the use of bicycles. The United Kingdom has built an 8,000-kilometer National Cycle Network that is scheduled to open in June 2000. This will pass within 4 kilometers of half the country's population, making it highly accessible and an obvious inducement to people to shift from cars to bicycles on short trips for shopping, commuting, and recreational riding.

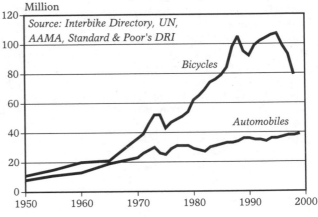

Figure 3: World Automobile and Bicycle Production, 1950–99

THE GLOBALIZATION OF INFORMATION

In 1998, the number of fixed-line phone connections worldwide reached 844 million, a gain of 7 percent over the preceding year. (See pages 92–93.) The number of cellular phone subscribers increased to 319 million, a phenomenal jump of 48 percent over the preceding year. In 1996, the number of new mobile phones exceeded the number of new fixed-line installations for the first time. Well before the end of this decade, the total number of cellular phones in use is likely to surpass the number of fixed-line phones.

Growth in the Internet has been even more impressive. By the end of 1999, some 72 million host computers were linked to the Internet, marking an expansion of 67 percent over 1998. (See pages 94–95.) They enabled 260 million people to go online. Although the U.S. share of this total is declining steadily, the United States—with 111 million Internet users—still accounts for 43 percent of the world total. The next four countries are Japan, with 18 million Internet users, and Canada, Germany, and the United Kingdom, with 14 million each.

Perhaps the most spectacular growth in 1999 occurred in developing countries, like Brazil (7 million users), China, and South Korea (about 6 million each). Internet access in developing countries nearly doubled in 1999. China alone expanded its access by fourfold, exceeding all projections.

The Internet has a huge potential for saving resources. Worldwide e-commerce in 1999 totaled $111 billion, triple the level for 1998. With the growing potential of ordering products from home and also working from home, one study projects that the U.S. area for malls and office buildings will be reduced by 3 billion square feet, greatly lowering use of both materials and energy.

Another environmental benefit of new technologies involves the use of satellites to monitor changes in Earth's physical condition. (See pages 140–41.) In 1997, for example, when fires were burning out of control during an intense drought in Indonesia, bringing air travel to a halt and making millions of people in the region physically sick, satellite images revealed the cause. The fires were concentrated in areas where plantation owners wanted to clear land for additional palm oil plantings; the fires that were burning out of control were not all accidental.

Some of the most helpful Earth-monitoring satellites are those used for weather. The science of weather forecasting and analysis has come a long way since the first weather satellite was launched in 1960. Today World Weather Watch, operated by the World Meteorological Organization, combines satellite observations with readings at ground, sea, and air monitoring stations, telecommunications links, and computer analysis centers to provide a highly sophisticated analysis and

short-term forecast of weather trends.

Satellites are playing an important role in measuring changes in snow cover as temperatures rise. They are also ideally situated to monitor the breakup of major ice sheets, such as those in West Antarctica. They can chronicle with great detail the shrinkage of the Aral Sea in Central Asia or Lake Chad in Africa. By recording fires in the Amazon, floods in China, and dust storms in the Sahara, they help us to monitor Earth's health.

SOCIAL TRENDS GRIM

During 1999, our numbers increased by 77 million, bringing world population to 6 billion. (See pages 98–99.) India, meanwhile, logged a demographic milestone of its own, surpassing 1 billion and joining China in the 1-billion club.

Nearly all the 77 million added to world population in 1999 were born in developing countries. Despite the obvious urgency of slowing population growth everywhere, some 120 million women have no access to family planning services at all. Another 350 million women in developing countries still lack convenient, regular access to safe family planning services.

The annual rate of world population growth, which dropped from more than 2 percent a generation ago to 1.3 percent in 1999, is now slowing in part because of rising mortality from the HIV epidemic. In 1999, 5.8 million people were infected with HIV, raising the total number infected to date to 49.9 million in 1999. (See pages 100–01.) AIDS deaths, which lag behind new infections by roughly eight years, totaled 2.6 million in 1999, up from 2.4 million the year before. An estimated 23 million Africans entered the new century with a death sentence imposed by the virus.

In a number of countries, including Botswana, Namibia, South Africa, and Zimbabwe, one fifth to one third of the adult population is HIV-positive. Unless there is a medical miracle, these countries will lose this huge segment of their adult population well before the end of this decade. Life expectancy is dropping precipitously in southern Africa, where expected life span had climbed from 44 in the early 1950s to 59 in the early 1990s, but is now expected to drop back to 45 during this decade. By 2010, AIDS orphans in Africa are expected to total 40 million, creating a new subclass and a massive social challenge.

After tracking HIV infection rates, mortality rates, and life expectancy changes associated with the disease, we are now beginning to see some of the secondary effects of the epidemic. In South Africa, more than 60 percent of the beds in some hospitals are occupied by AIDS patients, impairing the capacity of the health care system to satisfy basic heath care needs. Education, too, is being affected. While Zambia last year graduated 300 new teachers, AIDS claimed the lives of 600 teachers.

For some corporations operating in countries like Zimbabwe, Botswana, and South Africa, the cost of employee health insurance has doubled, tripled, or quadrupled over the last decade as the number of employees with AIDS has soared. The combination of declining life expectancy, falling investment levels, and the loss of a large share of the productive segment of the population to the virus is undermining the economic future of Africa.

Closely related to the spread of HIV is the dramatic resurgence in tuberculosis (TB) worldwide. (See pages 148–49.) In 1998, 8 million new TB cases were recorded, 95 percent of them in developing countries. Swelling populations of HIV-positive individuals with impaired immune systems provide a fertile ground for the TB virus to spread. And like other infectious diseases, TB moves rapidly around the world in an age of air travel.

A third epidemic, cigarette smoking in developing countries, is also measurably reducing life expectancy. Worldwide, the number of deaths from smoking-related causes is projected by the World Health Organization to increase from 4 million in 1998 to 10 million in 2030. (See pages 106–07.) Between now and 2015, cigarettes are projected to claim more lives than World War II did.

As public awareness of the social toll of cigarettes spreads, opposition to smoking is also growing. A movement that has gained great momentum in the United States is now spreading into Europe and many developing countries. In the United States, per capita cigarette consumption dropped by a record 9 percent in 1999 as a result of stiffer taxes, higher prices, and increased awareness of the health risks of smoking. The number of cigarettes smoked per person in the United States has dropped from about 2,875 in 1980 to 1,634 in 1999, a fall of 43 percent. In Europe, the number of smokers has dropped by 10 percent over the last decade. The European Union has banned all cigarette advertising after 2005.

Worldwide, cigarette consumption per person dropped from the historical high of 1,027 in 1990 to 915 in 1999, a drop of 11 percent. The World Health Organization has launched a major worldwide campaign to restrict cigarette smoking and to reduce the health toll associated with this often lethal habit.

The spread of the HIV virus, the resurgence of tuberculosis, and the increase in cigarette smoking in developing countries together may well reverse the steady worldwide rise in life expectancy that characterized the last half of the twentieth century.

Besides the additional crowding associated with the sheer growth in human numbers, the growing concentration of the world's population in cities is creating conditions that are conducive to the spread of infectious diseases. The world's urban population is growing at nearly 60 million per year, driven by migration from the countryside, by the natural increase within existing urban populations, and the absorption of villages by expanding cities. (See pages 104–05.)

As of 1999, 47 percent of the world's people lived in cities. By 2006, according to U.N. projections, more than half will live in cities, making humans for the first time in our existence a primarily urban species.

On the positive side, the number of people officially classified as refugees by the United Nations is declining. Between 1995, the historical high, and 1999, the number of refugees declined from 27.4 million to 21.5 million, a drop of 22 percent. (See pages 102–03.)

WARS AND PEACEKEEPING BOTH INCREASING

Last year the number of wars increased to 35, up from 32 the previous year. (See pages 110–11.) Among the eight new conflicts that broke out in 1999, two—those in Chechnya and East Timor—were widely covered in the media. Others, including conflicts in Tripura in Eastern India, Krygystan, and Nigeria, received little attention.

The overwhelming majority of wars in the world today are internal conflicts that are ethnic, religious, or tribal in nature and that are sometimes exacerbated by environmental degradation. Among the few international conflicts are the one between Ethiopia and Eritrea and the clashes between India and Pakistan. In human lives lost, the civil wars taking the greatest toll are the long-running wars in Afghanistan, claiming 1.9 million thus far, and the Sudan, 1.5 million.

Just as the increase in wars reversed a decline that had been under way for several years, so too an increase in peacekeeping expenditures in 1999 represented a reversal of a decline that had been going for four years. In 1999, estimated peacekeeping expenditures exceeded $1.4 billion, up from $860 million in 1998. (See pages 112–13.) At the end of 1999, some 14,600 soldiers, military observers, and civilian police drawn from 84 countries were serving in peacekeeping missions.

U.N. peacekeeping operations are still hamstrung by lagging financial support by the members of the United Nations. At the end of 1999, U.N. members were in arrears by a total of $1.7 billion, of which more than $1 billion was due from just one country—the United States.

ENVIRONMENTAL DETERIORATION

In recent years we have advanced our understanding of some of the effects of our chemi-

cal-dependent, throwaway economy on our natural environment. One of the more disturbing findings is the rise in pollutants in underground water supplies. Among the principal pollutants now widely found are pesticides, nitrates, petrochemicals, chlorinated solvents, heavy metals, and radioactive waste. (See pages 124–25.) Once long-lived pollutants make their way into the underground water supply, the damage is virtually irreversible. The health of hundreds of millions of people is now being affected by one or more of these water pollutants.

One particularly disturbing group of chemicals is the persistent organic pollutants, which have the potential to mimic the hormones that control reproduction, metabolism, and the functioning of immune systems. These disruptions appear to be affecting reproductive capacity in a number of species, including humans. (See pages 130–31.)

Another indicator of a deterioration of Earth's environment is the decline in various types of amphibians—frogs, toads, and salamanders. (See pages 128–29.) Evidence that amphibian populations were disappearing initially surfaced at the first World Congress of Herpetology in Canterbury, England, in 1989. At that time, it was thought that the observed declines might be the result of natural fluctuations. Today there is evidence worldwide that amphibian populations are indeed declining and disappearing. Among the apparent contributing factors are the clearcutting of forests, the loss of wetlands, the introduction of alien species, changes in climate, increased ultraviolet radiation, acid rain, and pollution from agriculture and industry.

In some situations, the immune systems of amphibians are weakened as a result of climate change or increased ultraviolet radiation, leaving them vulnerable to infectious diseases. Amphibians are particularly sensitive to change because they spend their lives in both aquatic and terrestrial environments and are affected by changes in both. In this sense, they are one of the most sensitive barometers of Earth's changing physical condition.

One way to reduce the pollutant load on the environment is to increase the recycling of materials, such as steel, tin, aluminum, plastic, and paper. For example, over the last quarter-century the amount of recovered paper has more than tripled, going from 35 million to nearly 110 million tons. (See pages 132–33.) But because the amount of paper used has increased so rapidly, the share of paper that is recycled increased only from 38 percent to 43 percent during this period. Nonetheless, this gradual rise is helping to reduce the pressure on forests and waste disposal systems, and is reducing both energy use and pollution.

Despite the emergence of the computer age and the "paperless office," world paper use continues to climb, increasing nearly 2 percent in 1998. (See pages 78–79.) Since 1950, paper use has increased sixfold, closely paralleling growth in the world economy. While 10 percent of the paper used worldwide goes into long-lasting products like books, the other 90 percent is used once as newspapers, packaging, or writing paper and then discarded. Among leading industrial countries, paper recycling rates range from a high of 72 percent in Germany to a low of 31 percent in Italy.

One social indicator of the response to environmental threats is the number of environmental treaties forged at the international level. Five new environmental agreements reached in 1999 brought this total to nearly 240. (See pages 134–35.) Some treaties are regional, others are global. They may be broad, focused on reducing carbon emissions or chlorofluorocarbon manufacturing worldwide, or they may be more narrow—devoted, for example, to reducing sulfur emissions in Europe or to managing a shared river system among neighboring countries in the Middle East.

Reaching agreement at the international level and signing a treaty is only the beginning. The treaties must then be enforced. Here the international community's performance is mixed. One of the most clearcut successes began with the Montreal Protocol in 1987, which initiated the phaseout of chlorofluorocarbons, the family of chemicals that is depleting the stratospheric ozone layer.

The environmental diplomacy that leads to the drafting and adoption of international treaties is emerging as a major component of international diplomacy. In many cases, it now supersedes the traditionally dominant activities, such as diplomacy related to military security issues. Among other things, this shift reflects the realization that threats to future political stability are becoming more environmental and less military in nature.

TAX SHIFTING TO SAVE THE ENVIRONMENT

As environmental threats have multiplied, environmentalists and political leaders have looked for ways to reverse the trends that are undermining our future. By far the most promising of these is shifting taxes from personal and corporate income to environmentally destructive activities, such as carbon emissions, the generation of toxic waste, the use of pesticides, and the use of virgin raw materials as opposed to recycled materials.

Sweden, starting in 1991, began shifting some of the tax burden from income to taxes on carbon and sulfur dioxide emissions. (See pages 138–39.) In the mid-1990s, several other countries followed suit, including Denmark, Finland, and the Netherlands. More recently, a second surge in tax shifting has occurred in Europe's largest industrial countries, including France, Germany, Italy, and the United Kingdom.

In some ways, the most dramatic environmental "tax" was introduced in the United States when the tobacco industry agreed to reimburse the 50 state governments with $251 billion for smoking-related health care expenditures incurred in the past. This sum, nearly $1,000 for each man, woman, and child in the country, in effect is a retroactive tax on cigarettes. In agreeing to this settlement, the tobacco industry implicitly accepted the principle that manufacturers are responsible for the indirect as well as the direct effects of using their products. In order to pay this enormous sum, the cigarette companies are forced to boost the prices of their cigarettes sharply, thus further discouraging consumption.

The great advantage of tax shifting is that it is far less cumbersome than regulation, permitting the market to continue to operate, thus exploiting its inherent efficiency. But by discouraging investments in environmentally destructive activities, such as coal burning, and encouraging investment in environmentally benign activities, such as wind electric generation, tax shifting steers the economy in an environmentally sustainable direction.

Part **ONE**

Key Indicators

Food

Trends

Grain Harvest Falls
Lester R. Brown

World grain production in 1999 fell to 1,855 million tons, down 1 percent from the 1,871-million-ton harvest of the year before.[1] (See Figure 1.) The fall in the 1999 harvest marked the second consecutive annual drop from the all-time high of 1,879 million tons reached in 1997.[2] In per capita terms, production declined to 309 kilograms in 1999, a fall of some 10 percent from the historical high of 342 kilograms in 1984.[3] (See Figure 2.)

Among the big three grains—wheat, rice, and corn—production of wheat and corn each fell by nearly 1 percent, while that of rice rose by just over 1 percent.[4] (See Figure 3.) At 598 million tons in 1999, the corn harvest maintained its historically recent edge over wheat, which came in at 584 million tons.[5]

China maintains its position as the world's leading grain producer: its harvest of 395 million tons exceeded the 333-million-ton harvest in the United States by some 19 percent.[6] India, with a harvest of 185 million tons, ranked third.[7] Combined, these three countries account for roughly half of the world grain harvest.[8]

The share of the world grain harvest used for feed remained essentially unchanged in 1999 at 37 percent.[9] Stated otherwise, more than one third of the world grain harvest is consumed indirectly in the form of livestock products. Among the individual grains, almost the entire rice harvest is consumed directly as food. By contrast, though corn is a food staple in many countries in Latin America and sub-Saharan Africa, worldwide it is used largely as feed. Consumption of wheat is more evenly divided between food and feed. It is the dominant food staple in the west, and also a leading staple in China and India. In Western Europe, Eastern Europe, and the former Soviet Union, wheat is also widely used for feed.

Perhaps the most interesting contrast in grain trends during the decade just ended was that between the former Soviet Union and China. Grain output in the former Soviet Union was in a free-fall during the 1990s.[10] Wheat production, for example, dropped from 102 million tons in 1990 to 66 million tons in 1999, a decline of one third.[11] Meanwhile, the coarse grain harvest dropped from roughly 103 million tons to 44 million tons, a staggering reduction of well over half, marking the first time in the modern era that a major industrial society has experienced such a sustained decline in food production.[12]

In China, by contrast, grain output during the 1990s went up by some 15 percent, climbing from 343 million to 395 million tons.[13] Few could have anticipated 20 years ago, or perhaps even 10 years ago, that the economic fortunes of the two communist giants would diverge so sharply during the 1990s. While China is emerging as an economic superpower, most of the 17 republics of the former Soviet Union are deteriorating economically. There is no indication that the worsening state of agriculture in Russia, the largest republic, will be reversed in the near future. The combination of political paralysis, corruption, and inept leadership appears likely to continue for some time.

Neither overall production nor world grain trade patterns have changed much in the last two years. Over the last four years, world wheat trade has fluctuated between 118 million and 125 million tons.[14] Trade in coarse grains, meanwhile, has remained steady at around 105 million tons, except in 1997 when higher prices cut it to roughly 100 million tons.[15] The international flow of rice, which increased from 20 million tons in 1996 to 27 million tons in 1997, has declined somewhat since then.[16]

With two consecutive declines in the world grain harvest, world carryover stocks of grain (the amount in the bin when the new harvest begins) in 2000 total some 66 days.[17] Although this is well above the all-time low of 53 days in 1996, it is still below the 70 days needed to cushion a poor harvest.[18] If the global economy expands by 3.5 percent, as projected, and world population increases by nearly 80 million, world demand for grain will climb during 2000.[19] Unless production rises accordingly, the weak grain prices of the late 1990s will start to recover.

WORLD GRAIN PRODUCTION, 1950–99

YEAR	TOTAL (mill. tons)	PER PERSON (kilograms)
1950	631	247
1955	759	273
1960	824	271
1965	905	270
1970	1,079	291
1971	1,177	311
1972	1,141	295
1973	1,253	318
1974	1,204	300
1975	1,237	303
1976	1,342	323
1977	1,319	312
1978	1,445	336
1979	1,410	322
1980	1,430	321
1981	1,482	327
1982	1,533	333
1983	1,469	313
1984	1,632	342
1985	1,647	339
1986	1,665	337
1987	1,598	318
1988	1,549	304
1989	1,671	322
1990	1,769	335
1991	1,708	319
1992	1,790	329
1993	1,713	310
1994	1,760	314
1995	1,713	301
1996	1,871	325
1997	1,879	322
1998	1,871	316
1999 (prel)	1,855	309

SOURCES: USDA, *Production, Supply, and Distribution,* electronic database, February 2000; USDA, "World Grain Database," unpublished printout, 1991; USDA, FAS, *Grain: World Markets and Trade*, February 2000.

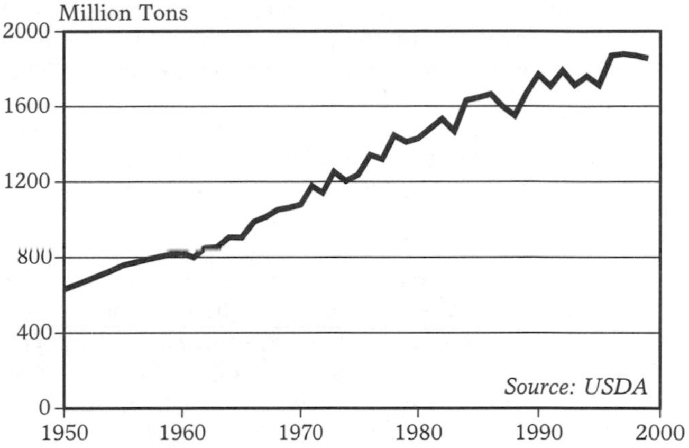

Figure 1: World Grain Production, 1950–99

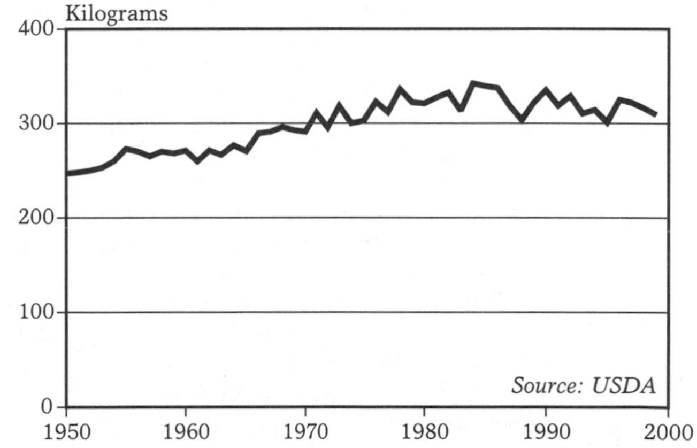

Figure 2: World Grain Production Per Person, 1950–99

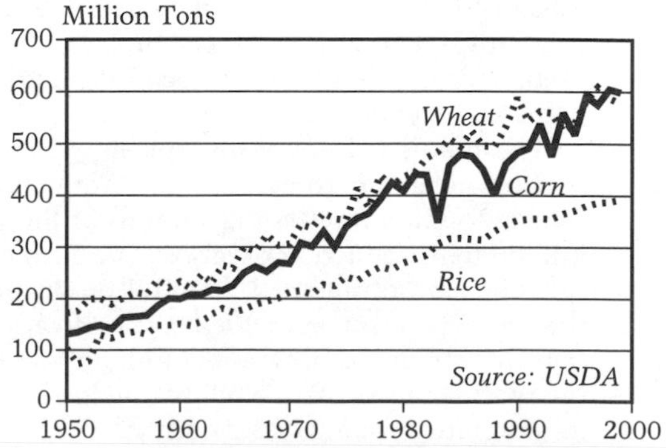

Figure 3: Wheat, Corn, and Rice Production, 1950–99

Soybean Harvest Drops   Lester R. Brown

The world soybean harvest in 1999 totaled 154 million tons, down 3 percent from the all-time record high of 159 million tons in 1998.[1] (See Figure 1.) Per capita production dropped from 26.9 to 25.6 kilograms, or 5 percent.[2] (See Figure 2.) The decline in production reflected adverse weather in some countries and lower prices from a weakening in overall demand caused by the economic disruptions in East Asia over the last couple of years.[3]

Over the last half-century, the production of soybeans has expanded faster than that of any other major crop, climbing from 17 million tons in 1950 to 154 million tons in 1999, a ninefold increase.[4] This compares with a tripling in the world grain harvest during the same period.[5]

The soybean, originally domesticated by early farmers in central China some 5,000 years ago, has come into its own during the last 50 years. In the United States, the harvested area of soybeans in 1999 was greater than that of any other crop, including wheat and corn—the traditional leaders.[6]

Demand for soybeans is closely tied to rising affluence. As incomes rise above the subsistence level, consumers everywhere begin to move up the food chain, consuming more animal protein in the form of meat, eggs, and milk. Production of poultry, eggs, and pork depends heavily on the use of soybean meal as a protein supplement to grain in feed rations.

In the world oilseeds economy, which supplies both vegetable oil and oil meal, soybeans dominate, accounting for 154 million tons of the 296-million-ton harvest in 1999.[7] (The other half consists of peanuts, sunflower seed, cottonseed, rapeseed, coconuts, and oil palm kernels.)[8]

When crushed, soybeans typically yield 68 percent meal and 16 percent oil.[9] Worldwide, they accounted for 104 million tons of the 166-million-ton production of oilseed meal, roughly 63 percent of the total.[10] For oil production, the figures are somewhat less impressive, with the soybean accounting for 24 million tons of the worldwide vegetable oil production of 85 million tons—roughly 28 percent.[11]

World production of soybeans is more concentrated than that of any other major crop: the United States, Brazil, Argentina, and China account for nearly 90 percent of the harvest.[12] The United States accounts for roughly half of the total, Brazil roughly a fifth, and Argentina and China about one tenth each.[13]

Within the United States, most of the soybeans are produced in the Corn Belt, often in an alternate-year rotation with corn. This helps control insects and diseases of both crops, and since the soybean is a legume, it fixes nitrogen—a nutrient for which the corn plant has a ravenous appetite. Today the Corn Belt is really the Corn-Soybean Belt.

Since 1950, the area planted to soybeans has grown from 14 million to 71 million hectares.[14] This fivefold expansion accounts for just over half of the growth in harvest, with the remainder coming from rising yield.[15]

Some countries, such as the United States, export soybeans largely as whole beans. Indeed, the United States accounts for 24 million tons of world soybean exports of 41 million tons.[16] The principal importers are the European Union, Japan, and China.[17]

Brazil and Argentina, the second and third ranking producers, crush most of their soybeans before exporting them as meal and oil. This helps explain why Argentina and Brazil dominate world soybean meal exports.[18] The leading importers are the European Union, which gets half of world soybean meal imports, and East Asia, particularly China and Japan, which takes much of the remainder.[19] Not surprisingly, Argentina and Brazil lead in oil exports as well, accounting for some 60 percent of the total.[20] Among the leading importers of soybean oil are China and India.[21]

If the global economy continues to expand and incomes continue to rise, particularly in low- and middle-income countries, the demand for the soybean either as meal or as oil is certain to increase. It also seems likely that the share of the world soybean harvest consumed directly as food, now less than one tenth, will expand in the years ahead as soybean products such as tofu compete with animal protein, such as meat and eggs, for a place in the human diet.[22]

WORLD SOYBEAN PRODUCTION, 1950–99

YEAR	TOTAL (mill. tons)	PER PERSON (kilograms)
1950	17	6.5
1955	19	7.0
1960	25	8.2
1965	32	9.5
1970	44	11.9
1971	47	12.5
1972	49	12.7
1973	62	15.9
1974	55	13.6
1975	66	16.1
1976	59	14.3
1977	72	17.1
1978	78	18.0
1979	94	21.4
1980	81	18.2
1981	86	19.0
1982	94	20.3
1983	83	17.7
1984	93	19.5
1985	97	20.0
1986	98	19.9
1987	104	20.6
1988	96	18.8
1989	107	20.7
1990	104	19.7
1991	107	20.0
1992	117	21.6
1993	118	21.3
1994	138	24.6
1995	125	22.0
1996	132	22.9
1997	158	27.1
1998	159	26.9
1999 (prel)	154	25.6

SOURCES: USDA, *Production, Supply, and Distribution*, electronic database, February 2000; USDA, FAS, *Oilseeds: World Markets and Trade*, February 2000.

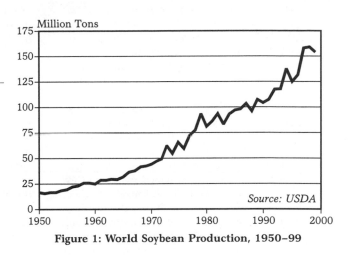

Figure 1: World Soybean Production, 1950–99

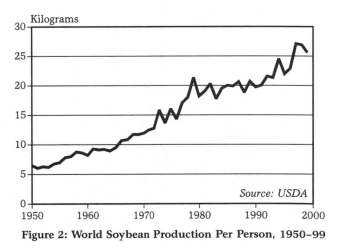

Figure 2: World Soybean Production Per Person, 1950–99

Meat Production Up Again Lester R. Brown

World meat production in 1999 totaled 217 million tons, up from 215 million tons in 1998—a gain of just 1 percent.[1] (See Figure 1.) With production lagging population growth ever so slightly, output per person fell from 36.4 kilograms in 1998 to 36.3 in 1999. (See Figure 2.)

The annual rise in meat production has become one of the most predictable trends in the world economy, increasing in each of the 39 years since 1960.

Production of beef, which increased little during the 1990s, maintained its slow growth at less than half of 1 percent in 1999.[2] (See Figure 3.) In the United States, the leading producer, it was up 2 percent from 11.8 million to slightly over 12 million tons.[3] In Brazil, the second ranking producer, output rose from 6.1 million to 6.3 million tons, a gain of over 3 percent.[4] In China, where beef production is being encouraged by the government, output was up nearly 2 percent.[5]

In two nations with export-oriented beef industries, Argentina and Australia, output was up 8 percent in the former and down 5 percent in the latter.[6] Production in Russia, which has been in an economic free-fall since 1990, dropped 9 percent from the preceding year.[7]

Production of pork, the world's leading source of meat, was up by less than 1 percent in 1999, climbing from 87.8 million to 88.3 million tons.[8] Production in China, which totally dominates the world pork economy, was up by roughly 2 percent, reaching 37 million tons.[9] In both the United States and the European Union, the other two big producers, it was up by 1 percent.[10] China and the United States (which produces 9 million tons) together account for half of the world's pork supply.[11]

World mutton production, which is a distant fourth on the meat production chart at scarcely 11 million tons, declined slightly in 1999.[12] Mutton production is concentrated in China, which accounts for a fourth of world consumption, and in Australia and New Zealand.[13] Annual consumption per person in New Zealand leads the world, at 29 kilo-grams, followed by two other mutton exporters—Australia at 18 kilograms and Ireland at 9 kilograms.[14] Affluent Saudi Arabia, heavily dependent on imported mutton, consumes 12 kilograms per person.[15]

World poultry production was up by nearly 3 percent in 1999, continuing to expand more rapidly than any other meat.[16] In the United States, the leading producer, output was up by nearly 6 percent.[17] In second-ranking China, growth slowed to less than 2 percent.[18] In Brazil, the number three producer, output was up by some 10 percent.[19] The three leading producers—the United States at 16 million tons, China at 12 million tons, and Brazil at 5 million tons—account for over half of world poultry production.[20]

World meat production increased from 44 million tons in 1950 to 217 million tons in 1999, gaining fivefold.[21] Expanding at roughly twice the rate of population, this more than doubled the meat produced per person.[22]

Accompanying this rapid growth was a dramatic shift in the pattern of world meat output. In 1950, beef was the leading source of meat, at 19 million tons, with pork following at 16 million tons, mutton a distant third at 5 million tons, and poultry at 4 million tons.[23] Today, pork has emerged as the leader largely because of the strong gains in output in China. Poultry has moved into second place; over-taking beef in 1995, it has steadily widened its margin since then. Mutton production remains a distant fourth.

The world's leading consumers of meat today are China and the United States.[24] Twenty years ago, the United States led the world by a wide margin.[25] But after the economic reforms in China in 1978, the Chinese economy expanded fourfold within two decades, and meat production surged ahead.[26] By 1999, China was eating 55 million tons of meat compared with 34 million tons in the United States.[27] With meat production growing faster in China than in the United States, this margin could widen even more as the decade unfolds.[28]

WORLD MEAT PRODUCTION, 1950–99

YEAR	TOTAL (mill. tons)	PER PERSON (kilograms)
1950	44	17.2
1955	58	20.7
1960	64	21.0
1965	80	24.0
1970	96	26.0
1971	100	26.5
1972	103	26.7
1973	104	26.3
1974	109	27.2
1975	111	27.1
1976	114	27.3
1977	117	27.8
1978	122	28.4
1979	127	21.1
1980	132	29.5
1981	134	29.6
1982	135	29.3
1983	140	29.8
1984	144	30.1
1985	149	30.6
1986	154	31.2
1987	159	31.7
1988	165	32.4
1989	168	32.4
1990	174	32.9
1991	177	33.1
1992	181	33.2
1993	185	33.6
1994	192	34.2
1995	197	34.8
1996	200	34.7
1997	208	35.6
1998	215	36.4
1999 (prel)	217	36.3

SOURCES: FAO, *1948–1985 World Crop and Live-stock Statistics* (Rome: 1987); FAO, *FAOSTATS*, electronic database, updated 7 December 1999.

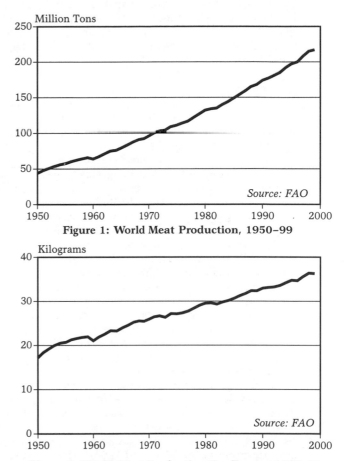

Figure 1: World Meat Production, 1950–99

Figure 2: World Meat Production Per Person, 1950–99

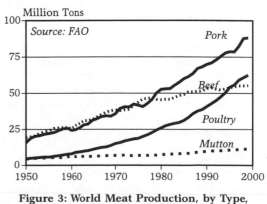

Figure 3: World Meat Production, by Type, 1950–99

Fish Harvest Down   Gary Gardner

Global fish catch was down sharply in 1998, the latest year with data available; it dropped by some 7.5 percent as unusual weather patterns reduced fish stocks in major fishing areas.[1] (See Figure 1.)

The decline in fish catch was partially offset by a robust 6.5-percent increase in output of farmed fish, as the aquaculture industry continued its rapid growth.[2] Overall, however, the depressed fish catch dominated fish supply in 1998, and total global supply fell 4.2 percent to 117 million tons.[3] Harvest per person fell 5 percent, to 19.8 kilograms.[4] (See Figure 2.)

The decline in fish catch resulted in part from the strongest El Niño weather event on record, which warmed the eastern Pacific in 1997–98 and reduced fish stocks.[5] Three of the world's five top producers—Peru, Chile, and the United States—all fish in waters affected by El Niño, and all saw declines in the catch.[6] China, on the other hand, the world's leading producer, saw fish catch increase in 1998 by a strong 9.6 percent.[7]

Fishing stress extends beyond areas affected by El Niño, however. The U.N. Food and Agriculture Organization (FAO) estimates that some 60 percent of the world's oceanic fisheries are fished at or beyond capacity.[8] Indeed, global fish catch grew by an average of more than 5 percent annually between 1950 and 1970, but then began to slow, dropping to just over 1 percent in the 1980s and 1990s.[9] The trend is worrisome in part because fish provides over 15 percent of humans' animal protein consumption, and about 6 percent of total protein consumption.[10]

Levels of fish catch are maintained in an increasingly depleted ocean in part by targeting smaller and smaller species. A 1999 report estimated that the average global marine catch in the past half-century has moved to a lower trophic level—a feeding scale that ranges from phytoplankton at the bottom to the largest species at the top.[11] Many of the smaller species prized today were considered inferior catch a few decades ago.

Fisheries are also depleted by high levels of bycatch, the nontargeted fish that turn up in nets. Some 20 million tons of bycatch—equal to nearly a quarter of the global fish catch—is captured each year, then thrown back to the sea, usually dead or dying.[12] The use of bycatch reduction devices is increasingly required in some countries to lessen the problem. Turtle excluder devices used in shrimping nets, for example, have meant a fourfold increase in turtle nests in the Gulf of Mexico since 1985.[13]

As fish catch has stalled over the past decade, aquaculture has picked up the slack, growing from some 8 percent of the world fish supply in 1984 to about 25 percent in 1998.[14] China is far and away the world's leading producer, accounting in 1998 for some two thirds of the global total.[15]

Aquaculture, however, often carries a stiff environmental and social toll. The feeding requirements of farmed salmon, shrimp, and other carnivorous species, for instance, often increase the pressure on fisheries to deliver more fish for use as feed.[16] Indeed, salmon farming can require two to four times more kilograms of fishmeal feed than it produces in salmon.[17]

Fish farming can also be highly polluting. The salmon fisheries of the Nordic countries release nitrogen in quantities equivalent to that found in the sewage of 3.9 million people, roughly the population of Norway.[18] This and other forms of pollution often limit the useful life of the fishery: most intensively cultivated shrimp ponds in Asia are used for no more than 5–10 years.[19] Aquaculture can be a source of biological pollution as well, when species foreign to a region escape into the wild and dominate ecosystems in which they have no natural enemies.

The future of global fisheries is uncertain. FAO projections of total fish harvest in 2010 range from 107 million to 144 million tons, depending on how fisheries and fish farms are managed.[20] The higher projection yields a fish availability per person that is barely above the 1998 level.[21] The lower projection would cut per capita levels by nearly a third, giving fish a much lower place in the diets of many in the future.[22]

WORLD FISH HARVEST, 1950–98

YEAR	WORLD CATCH (mill. tons)	AQUA-CULTURE (mill. tons)	HARVEST PER PERSON (kilograms)
1950	19		7.5
1955	26		9.4
1960	36		11.9
1965	49		14.7
1970	58		15.7
1971	62		16.4
1972	58		15.1
1973	59		15.0
1974	63		15.8
1975	62		15.2
1976	65		15.7
1977	63		14.9
1978	65		15.1
1979	66		15.1
1980	67		15.1
1981	69		15.3
1982	71		15.4
1983	72		15.4
1984	78	7	17.8
1985	79	8	17.9
1986	85	9	19.1
1987	85	10	19.0
1988	89	12	19.8
1989	89	12	19.6
1990	86	13	18.8
1991	84	14	18.3
1992	85	15	18.5
1993	86	18	18.9
1994	91	21	20.0
1995	92	24	20.5
1996	93	27	20.9
1997	93	29	20.9
1998	86	31	19.8

SOURCES: FAO, *Yearbook of Fishery Statistics: Capture Production* (various years); FAO, *Aquaculture Production* (various years); FAO, Fisheries Web site.

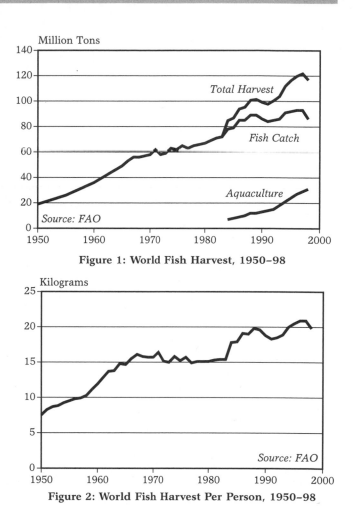

Figure 1: World Fish Harvest, 1950–98

Figure 2: World Fish Harvest Per Person, 1950–98

Agricultural Resource
Trends

Grain Area Shrinks Again   Gary Gardner

Grain harvested area fell to 674 million hectares in 1999, the smallest area since 1972.[1] (See Figure 1.) The reduction continues the general trend toward shrinkage that began in 1982. As global population increases, the harvested area per person continues to fall; it now stands at 0.11 hectares—more than a third smaller than in 1972.[2] (See Figure 2.)

Harvested area is the grain area that is reaped in a single year. (A hectare that is double-cropped in one year is counted as two hectares of harvested area.) Grains—principally corn, wheat, and rice—supply more than half the calories and protein eaten directly by humans.[3] Thus grain area tracks the resource base of the dominant component of the global food supply.

Wheat area fell by 3.6 percent while corn and rice area expanded marginally, by less than 1 percent each.[4] The trends in grain area swamped any increase in yields: corn and rice yields showed no increase, while a robust 3-percent increase in wheat yields was more than offset by the contraction in wheat area.[5]

Sometimes a contraction in grain area can be positive, as when growth in yields reduces the area needed to meet global grain demand, or when the area is reduced for fallowing, or when marginal land is converted from grain production to more sustainable uses such as pastureland. And grain area is often reduced simply because falling grain prices prompt some farmers to switch to more profitable crops. But reductions in area are problematic if yield growth is slow or nonexistent, as in 1999, or if land is lost permanently from cultivation.

Conversion of marginal land to more sustainable uses is now a common cause of cropland loss in some countries. The Chinese government is converting hillside farms and other marginal land to forests and pastureland; this laudable initiative is now the leading cause of cropland loss there.[6] In the United States, the Conservation Reserve Program (CRP) has removed some 12 million hectares of highly erodible land from production for 10–15 years to protect it and adjoining waterways from degradation.[7]

Urbanization is a small but growing threat to cropland in many countries. In China, construction of new roads, factories, and houses accounts for about one fifth of cropland loss.[8] In the United States, some 5.2 million hectares of cropland were lost between 1982 and 1997, while developed land expanded by 12 million hectares.[9] Most of the newly urbanized land was taken from forest or pastureland, but the trend is also of concern for cropland as urbanization rates increase.[10] Rates of urbanization in the United States doubled in the 1992–97 period compared with 1982–92.[11] Developed land now constitutes 7 percent of all U.S. non-Federal land, up from 5 percent in 1982.[12]

Degraded land is also removed from production, or loses productivity, in many parts of the world. In China, degradation—primarily erosion, but also desertification, salinization, and waterlogging—is responsible for as much cropland loss each year as urban and rural construction.[13] In the United States, the CRP reduced cropland erosion by 38 percent between 1985 and 1995, but erosion levels have plateaued since then.[14]

As area per person falls, countries turn increasingly to foreign markets for their grain. Japan, South Korea, and Taiwan, for example, now have less than a quarter of the world average grain area per person, and each imports more than 70 percent of its grain.[15] Population growth in many other Asian nations will reduce area per person to levels that have never supported food self-sufficiency anywhere. Indeed, by 2020 an estimated 70 percent of the people in Asia could depend on foreign markets for one fifth or more of their grain.[16]

Many see careful management of cropland as good for not only agriculture but the environment in general. Wetlands on farms often serve as stopovers for migratory birds, for example. And increasing soil carbon levels on farms could help sop up the excess carbon that is driving climate change. A 1999 study by the International Soil Reference and Information Centre estimated that 9–12 percent of human-produced carbon emissions could be absorbed by properly managed farms.[17]

WORLD GRAIN HARVESTED AREA, 1950–99

YEAR	AREA HARVESTED (mill. hectares)	AREA PER PERSON (hectares)
1950	587	0.23
1955	639	0.23
1960	639	0.21
1965	653	0.20
1970	663	0.18
1971	672	0.18
1972	661	0.17
1973	688	0.18
1974	691	0.17
1975	708	0.17
1976	716	0.17
1977	714	0.17
1978	713	0.17
1979	710	0.16
1980	722	0.16
1981	732	0.16
1982	716	0.16
1983	707	0.15
1984	711	0.15
1985	715	0.15
1986	709	0.14
1987	686	0.14
1988	688	0.14
1989	694	0.13
1990	694	0.13
1991	692	0.13
1992	693	0.13
1993	684	0.12
1994	684	0.12
1995	681	0.12
1996	702	0.12
1997	690	0.12
1998	686	0.12
1999 (prel)	674	0.11

SOURCE: USDA, *Production, Supply, and Distribution*, electronic database, February 2000.

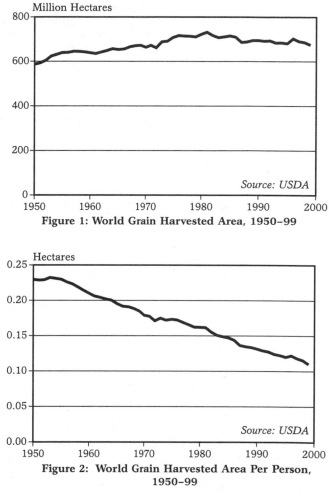

Figure 1: World Grain Harvested Area, 1950–99

Figure 2: World Grain Harvested Area Per Person, 1950–99

Fertilizer Use Down   Lester R. Brown

World fertilizer use in 1999 totaled 134 million tons, down from 137 million tons in 1998.[1] (See Figure 1.) Indeed, in each of the last three years, 1997 to 1999, fertilizer use has been essentially flat, fluctuating narrowly between 134 million and 137 million tons.[2] As a result, fertilizer use per person worldwide is slowly declining as population continues to grow by nearly 80 million per year.[3] In 1999, it was just above 22 kilograms per person, a drop of 21 percent from the peak of 28 kilograms in 1989.[4]

Fertilizer use has leveled off since 1997 largely because of disruptions in the global economy. The demand for agricultural commodities began to weaken after the Asian financial crisis began in July 1997 with the devaluation of the Thai baht.[5] This weakening was reinforced by the massive debt default in Russia in September 1998, and by Brazil's devaluation of its currency in January 1999.[6] With world grain prices in 1999 at their lowest level in two decades or so, there was little incentive for farmers to raise fertilizer use.[7]

Another reason fertilizer consumption is stagnant is the diminishing production response to additional usage in key countries. Among the countries or regions where fertilizer use has plateaued are the United States, Canada, Western Europe, Japan, Taiwan, and perhaps China. In contrast, usage is still growing vigorously in India and Brazil. Fertilizer applications in India leveled off in 1999, but had jumped by 13 percent in 1998.[8] And in Brazil, usage grew by 13 percent in 1998 and by 6 percent in 1999, one of the strongest growth trends in any country.[9]

Trends in the big four agricultural countries—China, India, the former Soviet Union, and the United States—show some sharp contrasts over the last two decades. Perhaps the most dramatic and unexpected change was the precipitous decline in fertilizer use in the Soviet Union after the economic decline that began a decade ago.[10] (See Figure 2.)

In China, on the other hand, fertilizer use soared after the economic reforms in 1978, climbing from some 6 million tons in 1977 to the all-time high of nearly 36 million tons in 1997.[11] Since then, it has fallen to 31 million tons.[12] Given the intensity of current fertilizer applications in China, it seems unlikely that usage will expand much in the future. Indeed, we may have witnessed a plateauing of fertilizer use in China in the late 1990s that is similar to the one that began in the United States in the early 1980s.[13]

In the United States, the leader in applying fertilizer throughout most of the third quarter of this century, usage hit an all-time high in 1980 of just over 21 million tons.[14] (See Figure 3.) Since then, it has averaged roughly 19 million tons a year.[15]

In India, which used hardly any fertilizer in 1960, consumption has increased rather steadily since 1975, climbing to 16 million tons in 1999.[16] It may increase somewhat further, but not a great deal since it is already approaching the amount used by U.S. farmers in 1999.[17]

Given the growing world demand for food, fertilizer use is likely to continue rising at the global level. But the growth rate will probably be modest simply because in more and more countries farmers are reaching the point where additional fertilizer use has little effect on production. Crops now in use are physiologically incapable of absorbing many more nutrients.

Another emerging constraint on the growth in world fertilizer demand is water scarcity, especially in China and India, the world's two most populous countries.[18] In both, farmers are losing irrigation water to cities and to aquifer depletion.

Further constraining fertilizer use is nutrient runoff. This is seen as a serious problem in Europe, where fertilizer use has declined somewhat in recent years, and in the United States, where fertilizer nutrients flowing down the Mississippi River and into the Gulf of Mexico are leading to explosions of algae.[19] When these algae concentrations die, they absorb the free oxygen in the water, leading to the death of all marine life in that area, including various types of seafood. In effect, efforts to expand the harvest from the land are reducing the harvest from the oceans.

WORLD FERTILIZER USE, 1950–99

YEAR	TOTAL (mill. tons)	PER PERSON (kilograms)
1950	14	5.5
1955	18	6.5
1960	27	8.9
1965	40	12.0
1970	66	17.8
1971	69	18.2
1972	73	18.9
1973	79	20.1
1974	85	21.2
1975	82	20.1
1976	90	21.6
1977	95	22.5
1978	100	23.2
1979	111	25.4
1980	112	25.1
1981	117	25.8
1982	115	24.9
1983	115	24.5
1984	126	26.4
1985	131	27.0
1986	129	26.2
1987	132	26.3
1988	140	27.4
1989	146	28.1
1990	143	27.1
1991	138	25.7
1992	134	24.6
1993	126	22.8
1994	121	21.6
1995	122	21.5
1996	129	22.4
1997	135	23.1
1998	137	23.1
1999 (prel)	134	22.3

SOURCES: FAO, *Fertilizer Yearbook* (Rome: various years); Soh and Isherwood, "Short Term Prospects for World Agriculture and Fertilizer use," IFA Meeting, 30 November–3 December 1999; Aholou-Putz, IFA, e-mail, 27 January 2000.

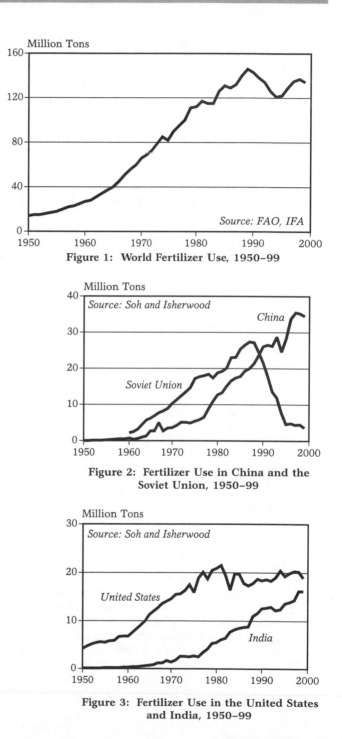

Figure 1: World Fertilizer Use, 1950–99

Figure 2: Fertilizer Use in China and the Soviet Union, 1950–99

Figure 3: Fertilizer Use in the United States and India, 1950–99

Pesticide Trade Nears New High — Sarah Porter

In 1998, world exports of pesticides stood at $11.4 billion (in 1998 dollars), nearly nine times the level in 1961.[1] (See Figure 1.) This is a 5.4-percent increase over 1997, when trade was adversely affected by the economic slump in Asia. Indeed, exports in 1998 were just below their peak of $11.7 billion in 1996.[2]

The last three decades have seen strong growth in pesticide trade, as pesticide-intensive farm practices spread throughout the industrial world and to many developing nations.[3] In the early 1980s, trade slowed as many countries went into recession. After resuming growth of almost 5 percent annually, trade slumped again in the early 1990s due to renewed recession as well as farmer uncertainty over agricultural reforms in Western Europe.[4]

Industrial-country exports accounted for 85 percent of the value of trade in 1998.[5] Western Europe and the United States exported $7.1 billion and $1.7 billion in pesticides respectively, representing 62 and 15 percent of the world trade total.[6] France and Germany retained their positions as the world's top two exporters.[7]

With 15 percent of world exports but 33 percent of world imports, developing countries had net imports of pesticides worth $2.2 billion in 1998.[8] (See Figure 2.) China is the largest exporter and importer in the group, and it ranks sixth worldwide in the value of its exports.[9] For the first time since 1973, however, the value of Latin American pesticide imports passed that of Asian developing countries in 1997 and 1998.[10] Argentina, Brazil, and Mexico are the second, third, and fourth largest developing-country importers.[11]

International trade of pesticides accounted for about 37 percent of an estimated $31 billion in world sales in 1998.[12] Based on value of sales, North America uses about 30 percent of the world's pesticides, while Western Europe uses 26 percent and East Asia 22–24 percent. Latin America accounts for about 11 percent of world pesticide sales, with Brazil being one of the top five users in the world. Africa uses some 4 percent.[13]

But a region's share of global pesticides sales may not correspond exactly with its share of pesticide usage. For instance, many farmers in industrial countries have been moving toward higher-value chemicals that are more pest-specific and used in lower doses than older pesticides. At the same time, cheaper, older, and higher-dose pesticides are the mainstays of farmers in many developing nations.[14]

Pesticide use per hectare has risen dramatically worldwide since 1961, from 0.49 kilograms per hectare to 1.79 kilograms in 1995.[15] (See Figure 3.) By the 1990s, pesticide use began leveling off in most industrial countries and is not expected to increase dramatically.[16] The agrochemical industry increasingly expects growth to come from developing nations.[17]

Poor weather conditions and pest outbreaks, changes in crop acreages, government regulations, and economic factors such as commodity prices all have an impact on pesticide usage.[18] Flooding in the U.S. Midwest in 1993 reduced usage, while a severe drought in 1997 caused pesticide use to fall in India and Thailand by more than 10 percent.[19] Other countries, such as Sweden and Denmark, are deliberately cutting pesticide use.[20] And in 1986, Indonesia began promoting integrated pest management, banning the import of 57 pesticides. The value of its pesticide imports dropped from $124.9 million in 1976 to $22.4 million in 1986. Since then, they have not risen above $37 million.[21]

At the international level, there has been a move to regulate pesticides with especially adverse effects on human and environmental health. In September 1998, negotiations concluded on the Rotterdam Convention on the Prior Informed Consent Procedure for Certain Hazardous Chemicals and Pesticides in International Trade.[22] This treaty requires importing countries to be told whether a chemical is banned or severely restricted in the exporting country and what its harmful effects are. By late 1999, 73 countries had signed the treaty.[23]

WORLD EXPORTS OF PESTICIDES, 1961–98

YEAR	EXPORTS (bill. 1998 dollars)
1961	1.3
1962	1.5
1963	1.6
1964	1.8
1965	1.6
1966	1.9
1967	2.0
1968	2.0
1969	2.2
1970	2.3
1971	2.5
1972	2.7
1973	3.6
1974	4.9
1975	5.4
1976	4.8
1977	5.5
1978	6.5
1979	7.0
1980	7.6
1981	6.5
1982	6.1
1983	6.3
1984	6.6
1985	6.4
1986	7.0
1987	7.8
1988	8.2
1989	8.6
1990	9.0
1991	8.5
1992	8.4
1993	8.4
1994	9.7
1995	11.0
1996	11.7
1997	10.8
1998	11.4

SOURCES: FAO, *FAOSTAT Statistics Database*, 21 December 1999.

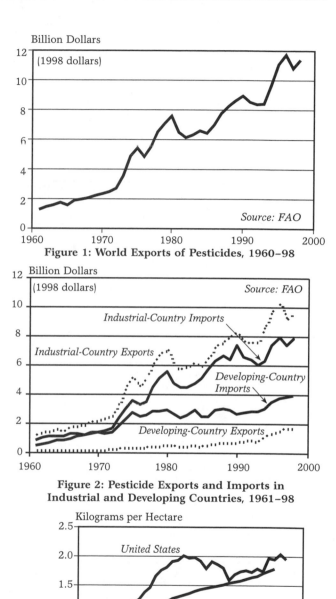

Figure 1: World Exports of Pesticides, 1960–98

Figure 2: Pesticide Exports and Imports in Industrial and Developing Countries, 1961–98

Figure 3: Pesticide Intensity in Agriculture, World, 1961–95, and the United States, 1964–97

Energy Trends

Fossil Fuel Use in Flux   Seth Dunn

In 1999, world consumption of fossil fuels increased 0.1 percent, to just above 1997 levels.[1] (See Figure 1.) Although usage expanded more than fourfold over the last 50 years, average annual growth rates slowed from 5.7 percent during the 1950s to 0.7 percent during the 1990s.[2] Divergent trends for individual fuels, meanwhile, reflect an ongoing shift in the global energy system from solid to liquids to gases.

Petroleum use rose 0.9 percent, bringing it to an average annual growth rate of 0.8 percent for the 1990s.[3] The greatest growth took place in the developing world, with a 3.5-percent increase in rebounding Asian economies.[4] The United States saw a 2.2-percent rise in oil use in 1999.[5]

Use of the fastest-growing fossil fuel, natural gas, rose 3.0 percent in 1999 and grew at an average annual rate of 1.9 percent in the 1990s.[6] Consumption in developing economies and Europe rose 5.9 and 4.7 percent, respectively, driven by demand for gas-based power generation and heating.[7] The latest estimates indicate that natural gas use passed coal use in 1998 for the first time.[8]

Coal consumption dropped 3.3 percent in 1999—reaching its lowest level since 1986.[9] Global coal use experienced negative growth during the 1990s.[10] Consumption fell 3 percent in China in 1999, 1.7 percent in the United States, 4.5 percent in Europe, and 7.3 percent in former Eastern bloc nations.[11]

Coal's decline may steepen in coming years, with important social effects. In the United Kingdom, two thirds of the remaining deep coal mines reportedly face closure this year in the absence of official aid.[12] According to an independent study by CLG Energy Consultants, U.K. government efforts to slow the decline of the coal industry—restricting construction of gas-fired power stations, aiming to halt production subsidies in Germany and Spain—have misfired.[13] In the United States, evidence of persistent acid rain and smog problems has prompted lawsuits from the New York State Attorney General and the Environmental Protection Agency against utilities upgrading decades-old power plants

to increase output without updating their pollution controls, as clean air regulations require.[14]

Analysts at the Lawrence Berkeley National Laboratory expect China's slowdown in coal use to continue during 2000 as the government closes down more state-owned mines and as homeowners shift to natural gas and other alternative energy sources for residential heating and cooking.[15] Indeed, China illustrates how economic and environmental pressures increasingly favor natural gas. The government approved plans for importing natural gas in 1999 and in April 2000 will begin constructing its first major natural gas pipeline, between the northwest and polluted Lanzhou.[16] It is also working with Russia to build pipelines between Siberian gas fields and northern cities such as Beijing and Tianjin.[17]

Recent fluctuations in oil prices—a drop in late 1998 to the lowest levels in 26 years, followed by a near tripling in 1999—are fueling debate over oil's future availability and use.[18] (See Figure 2.) Some experts argue that cheap oil is likely to return, with technological innovations constantly lowering the cost of discovering and exploiting new oil fields.[19] Amy Myers Jaffe and Robert Manning argue that the energy problem in the early twenty-first century will be a prolonged oil surplus and low oil prices.[20] Others note that the major oil discoveries of the 1970s—such as in Alaska and the North Sea—are reaching their limits, and that the proportion of oil reserves from outside the Middle East has not changed significantly since the 1970s, leaving the world vulnerable to another oil embargo.[21] Colin Campbell contends in *The Coming Oil Crisis* that oil production will peak and decline during this decade, causing local shortages and oil price shocks.[22]

The salient question, however, is not whether the world will run out of oil—or fossil fuels more generally—but how much more carbon dioxide from burning these fuels can be absorbed by the atmosphere before dangerous climatic disruptions take place.[23] The real danger is not running out of fossil fuels, but continuing to use them at unsustainable rates.

Fossil Fuel Use in Flux

World Fossil Fuel Consumption, 1950–99

YEAR	COAL	OIL	NATURAL GAS
	(mill. tons of oil equivalent)		
1950	1,043	436	187
1955	1,234	753	290
1960	1,500	1,020	444
1965	1,533	1,485	661
1970	1,635	2,189	1,022
1971	1,632	2,313	1,097
1972	1,629	2,487	1,150
1973	1,668	2,690	1,184
1974	1,691	2,650	1,212
1975	1,709	2,616	1,199
1976	1,787	2,781	1,261
1977	1,835	2,870	1,283
1978	1,870	2,962	1,334
1979	1,991	2,998	1,381
1980	2,021	2,873	1,406
1981	1,816	2,781	1,448
1982	1,878	2,656	1,448
1983	1,918	2,632	1,463
1984	2,001	2,670	1,577
1985	2,100	2,654	1,640
1986	2,135	2,743	1,653
1987	2,197	2,789	1,739
1988	2,242	2,872	1,828
1989	2,272	2,921	1,904
1990	2,245	2,964	1,942
1991	2,190	2,956	1,981
1992	2,172	2,978	1,990
1993	2,162	2,943	2,037
1994	2,174	2,998	2,049
1995	2,207	3,031	2,116
1996	2,285	3,106	2,213
1997	2,266	3,174	2,202
1998	2,219	3,171	2,234
1999 (prel)	2,146	3,200	2,301

SOURCE: Worldwatch estimates based on UN, BP, DOE, EC, Eurogas, PlanEcon, IMF, and LBL.

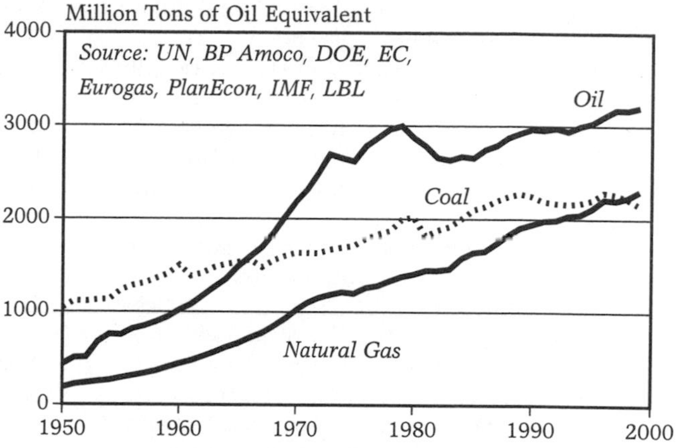

Figure 1: World Fossil Fuel Consumption, 1950–99

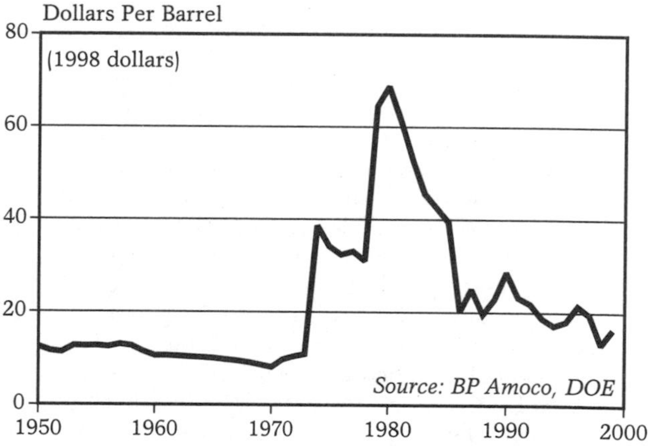

Figure 2: Real Price of Oil, 1950–99

Nuclear Power Rises Slightly   Nicholas Lenssen

Between 1998 and 1999, total installed world nuclear generating capacity increased 0.4 percent, or just 1,440 megawatts, to 344,526 megawatts.[1] (See Figure 1.) During the 1990s, global capacity rose only 4.7 percent, compared with total growth in the 1980s of 140 percent.[2] The all-time peak in global nuclear capacity will likely occur within the next three years.

Construction started on two reactors in 1999 (see Figure 2), both in Japan, bringing the total under construction to 32 reactors with a combined capacity of 25,716 megawatts.[3] This is the fewest number of reactors being built in more than 30 years.

Three new reactors—one each in France, India, and the Slovak Republic—were connected to the grid in 1999, and one reactor in Sweden was permanently closed.[4] Thus, 95 reactors have been retired after an average service life of less than 18 years, representing a generating capacity of 28,779 megawatts.[5] (See Figure 3.) By the end of 1999, 431 reactors were operating, one more than six years earlier.[6]

Not a single reactor is under construction in North America or Western Europe, and no new projects are expected for at least the next few years. The gradual opening of electricity markets in both regions is putting enormous pressures on nuclear operators to become more competitive. As much as 40 percent of the U.S. nuclear capacity is vulnerable to permanent shutdown due to high costs.[7] The U.S. Department of Energy estimates that 31 percent of the country's nuclear capacity will be closed by 2015.[8]

In Europe, the French government renewed its moratorium on new nuclear projects until after presidential elections in 2002.[9] At the same time, many countries are accelerating plans to permanently close their uneconomic and aging reactors. Sweden confirmed it would close a second reactor in 2001, the United Kingdom committed to closing two reactors in 2002, and the Netherlands announced it would proceed with plans to shut its last reactor in 2003.[10] The German government is still negotiating with the industry on a timetable for shutting down the country's 19 nuclear power reactors.[11]

Bulgaria has announced it will close two reactors in 2002 and another two by 2006.[12] Lithuania plans to close one by 2005, Russia one by 2004, and the Slovak Republic two reactors by 2008.[13] Kazakhstan has announced plans to close its sole reactor, which is located near the Iranian border.[14]

In 1999, Japan experienced the world's most serious nuclear accident since the 1986 Chernobyl explosion. The September accident at a nuclear fuel fabrication facility not far from Tokyo killed one worker and further jeopardized the government's plan to build 20 new reactors by 2010.[15] By the end of 1999, just four Japanese reactors were still under construction, and the head of the government energy advisory committee publicly questioned the credibility of the official nuclear plans.[16]

Elsewhere in Asia, China's government, which hopes to increase the country's nuclear capacity from the 6,500 megawatts now operating or under construction to 40,000 megawatts by 2020, unexpectedly announced in 1999 that it would not sanction additional nuclear plants for at least three years, leaving plans to build two Russian-designed reactors in limbo.[17]

South Korea aims to add 14 new reactors, of which 6 are under construction, to its existing 14 by 2015.[18] And India, which plans to add another new, small reactor to the grid in 2000, aims to have 2,000 megawatts operating in 2000, up from 1,900 megawatts at the end of 1999.[19] These plans face formidable economic obstacles, however.

One glimmer of hope for the nuclear industry is Turkey, which appears to be close to ordering its first nuclear power plant after ineffectually pursuing it for some 30 years.[20] The plan has generated extensive opposition from nongovernmental organizations, however, particularly since a deadly earthquake rattled a neighboring region in 1998. As 1999 ended, the Turkish government postponed for a third time a decision on issuing an order.[21]

WORLD NET INSTALLED
ELECTRICAL GENERATING CAPACITY
OF NUCLEAR POWER PLANTS,
1960–99

YEAR	CAPACITY (gigawatts)
1960	1
1965	5
1970	16
1971	24
1972	32
1973	45
1974	61
1975	71
1976	85
1977	99
1978	114
1979	121
1980	135
1981	155
1982	170
1983	189
1984	219
1985	250
1986	276
1987	297
1988	310
1989	320
1990	328
1991	325
1992	327
1993	336
1994	338
1995	340
1996	343
1997	343
1998	343
1999 (prel)	345

SOURCE: Worldwatch Institute database, compiled from the IAEA and press reports.

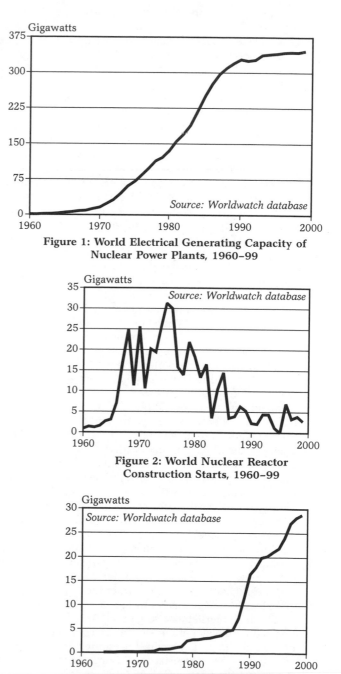

Figure 1: World Electrical Generating Capacity of Nuclear Power Plants, 1960–99

Figure 2: World Nuclear Reactor Construction Starts, 1960–99

Figure 3: Cumulative Generating Capacity of Closed Nuclear Power Plants, 1964–99

Wind Power Booms
Christopher Flavin

Wind power, already the world's fastest-growing energy source, surged to an even higher growth rate in 1999 as generating capacity rose 39 percent, to an estimated 13,840 megawatts.[1] (See Figure 1.) At the dawn of a new century, wind power provides eight times as much electricity to the world's consumers as it did just a decade earlier.[2]

The estimated 3,900 megawatts of wind turbines installed in 1999 is 65 percent higher than the year before (see Figure 2); the wind turbine market is now growing almost as quickly as the booming global market in mobile phones, which grew 73 percent in 1999.[3] The wind turbines installed in 1999 were worth over $3 billion, and supported roughly 86,000 jobs.[4]

For the sixth year in a row, Germany dominated global wind installations in 1999, with an estimated 1,568 megawatts installed, almost twice as much as in 1998.[5] (See Figure 3.) Germany's 4,445 megawatts of wind turbines now provide 2 percent of the country's electricity, and over 10 percent in some windy northern regions.[6] Wind development is spreading from the northwestern states where it started to less windy inland sites, as well as to the Baltic coast.

Spain surged to second place in new installations in 1999 at 750 megawatts, and to number four in total capacity, at 1,584 megawatts.[7] From Galicia in the northwest to Catalonia in the northeast and Andalusia in the south, Spain's wind industry shows every sign of even faster growth in the years immediately ahead. As the new year began, Energia Hidroelectrica de Navarra, Spain's leading wind energy developer, announced the largest single order for wind turbines ever—1,800 turbines with a generating capacity of roughly 1,400 megawatts—some $700 million worth of turbines.[8] (This is twice the size of the entire global wind power market in 1994.)[9]

The U.S. market also surged in 1999, boosted by developers' rush to take advantage of a wind energy tax credit that expired at the end of June (later reinstated by Congress at the end of the year).[10] The 562 megawatts added in 1999 took U.S. capacity to 2,490 megawatts—keeping it, for the moment at least, in the number two position in total installed capacity.[11] The largest new installations were in the country's heartland—Iowa, Minnesota, and Texas—where the wind potential is far larger than in California, where the U.S. industry got started in the 1980s.[12]

Other countries with growing wind power industries in 1999 were Denmark, with 290 new megawatts installed, Italy with 101 megawatts, and Greece with 73 megawatts.[13] In the developing world, wind development continues to lag due to a lack of policy support, but limited development is proceeding in China, India, and Costa Rica.[14]

As wind technology continues to advance, it is rapidly closing the cost gap with conventional power plants. The U.S. Department of Energy estimates that wind power now costs 4–6¢ per kilowatt-hour—about the same as for new gas- and coal-fired power plants.[15] One of the main factors driving costs down is the growing size of wind turbines, many of which now have blade spans of over 70 meters and generate between 1,000 and 2,000 kilowatts (1–2 megawatts) of electricity.[16]

Large-scale offshore wind development continues to move closer in Europe. Denmark, Germany, Ireland, the Netherlands, Sweden, and the United Kingdom are among the countries where major corporations like Royal Dutch Shell are making plans for massive wind projects in the North and Baltic Seas.[17] A study by Germanischer Lloyd and Garrad Hassan concluded that Europe's offshore wind potential in waters of 30 meters depth or less could supply all of the continent's power.[18]

A 1999 study sponsored by the European Wind Energy Association, the Forum for Energy and Development, and Greenpeace International estimated that if recent growth rates are sustained, wind power could supply 10 percent of the world's electricity by 2020.[19] For this to happen, annual investments will have to increase roughly to $78 billion in 2020, equivalent to about 40 percent of annual investments in all electric generating capacity in the 1990s.[20]

World Wind Energy Generating Capacity, Total and Annual Addition, 1980–99

YEAR	TOTAL	ANNUAL ADDITION
	(megawatts)	
1980	10	5
1981	25	15
1982	90	65
1983	210	120
1984	600	390
1985	1,020	420
1986	1,270	250
1987	1,450	180
1988	1,580	130
1989	1,730	150
1990	1,930	200
1991	2,170	240
1992	2,510	340
1993	2,990	480
1994	3,680	720
1995	4,820	1,294
1996	6,115	1,290
1997	7,640	1,566
1998	9,940	2,363
1999 (prel)	13,840	3,900

Sources: EWEA, AWEA, FGW, Jose Santamarta, and *Renewable Energy Report.*

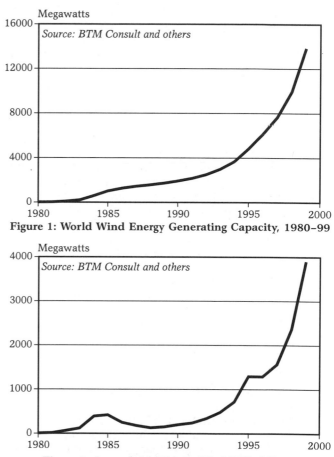

Figure 1: World Wind Energy Generating Capacity, 1980–99

Figure 2: Annual Addition to World Wind Energy Generating Capacity, 1980–99

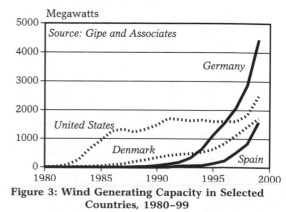

Figure 3: Wind Generating Capacity in Selected Countries, 1980–99

Solar Power Market Jumps   Christopher Flavin

Production of solar photovoltaic (PV) cells grew to an estimated 201.3 megawatts in 1999, up 30 percent from 1998.[1] (See Figure 1.) Production has grown almost fivefold in the past 10 years.[2] The wholesale market for solar modules was just under $700 million in 1999, while the total market for PV systems, including equipment such as batteries and inverters, as well as installation costs, was roughly $1.6 billion.[3]

The growth in the PV market in 1999 was propelled by two main uses for the technology: off-grid applications such as powering telecommunications, traffic signals, and village households; and grid-connected rooftops in industrial countries, which are supported by government subsidies in Europe and Japan.[4]

The Japanese PV industry surged into the lead in 1999 by producing 80 megawatts of cells, up 63 percent in just a year.[5] Most Japanese PVs went to the country's generously supported residential solar program, which resulted in over 9,000 PV systems being installed in 1999.[6] Japan's Kyocera is now the world's number two PV producer, at 30.3 megawatts in 1999.[7]

U.S. PV production grew more slowly, reaching 60.8 megawatts, much of which was exported, mainly to the developing world.[8] Although the Clinton administration launched an impressive-sounding Million Solar Roofs program, it provides less support than do most other countries with similar programs.[9] The U.S. Department of Energy announced an initiative aimed at deploying solar cells atop abandoned urban properties, but Congress has shown little interest in funding it adequately.[10]

California-based PV producer Siemens Solar has slipped in just two years from being the world's number one PV producer to number four.[11] Meanwhile, BP Solarex, the product of an oil industry merger, tops the list of PV producers, though its production is split between plants in the United States, Europe, and other locations.[12]

European PV production grew from 33.5 megawatts in 1998 to 40.0 megawatts in 1999, propelled by solar home programs in countries such as Germany, Norway, and Switzerland.[13] Further market growth in Europe is likely in the next few years. In November, the solar division of Royal Dutch Shell opened a large, 25-megawatt, fully automated solar cell production plant in Germany.[14]

Efforts to promote PVs in developing countries continued to pick up pace in 1999. Sporadic subsidies from governments and international agencies are the fuel that keeps most such efforts growing. For example, a "barefoot college" in India's Rajasthan state is working to train village women to manufacture and repair solar lanterns.[15] The South African government is planning to provide solar electricity for rural schools and clinics, and announced in 1999 that it will install 350,000 solar home systems.[16]

Polycrystalline silicon cells dominated the PV market in 1999, with a 44-percent share, followed by the traditional single crystal silicon cells, which held 36 percent.[17] Thin-film amorphous silicon accounted for 16 percent of the 1999 market, including some 8.6 megawatts used for indoor applications such as calculators.[18] Although amorphous cells are less efficient than crystalline cells, many analysts believe that they hold the best potential for cutting manufacturing costs dramatically in the years ahead.

Advancing technology and a competitive marketplace pushed solar module prices down sharply in 1999 to $3.50 per watt or $3,500 per kilowatt.[19] (See Figure 2.) This represents a breakthrough since PV prices had been "stuck" in the $4.10–$4.30 range for six years.[20] Further cost declines are needed to make solar electricity competitive in most grid-connected applications; if prices continue to decline, market growth should accelerate in the years ahead.

One promising indication of future trends was a dramatic surge in share prices for "alternative" energy technology companies in January 2000.[21] Among the solar companies that benefited from the jump in investor interest were AstroPower, Energy Conversion Devices, Spire, and SolarWorld.

Solar Power Market Jumps

World Photovoltaic Production, 1971–99

YEAR	PRODUCTION (megawatts)
1971	0.1
1975	1.8
1976	2.0
1977	2.2
1978	2.5
1979	4.0
1980	6.5
1981	7.8
1982	9.1
1983	17.2
1984	21.5
1985	22.8
1986	26.0
1987	29.2
1988	33.8
1989	40.2
1990	46.5
1991	55.4
1992	57.9
1993	60.1
1994	69.4
1995	78.6
1996	88.6
1997	125.8
1998	154.9
1999 (prel)	201.3

SOURCE: Paul Maycock, *PV News*, various issues.

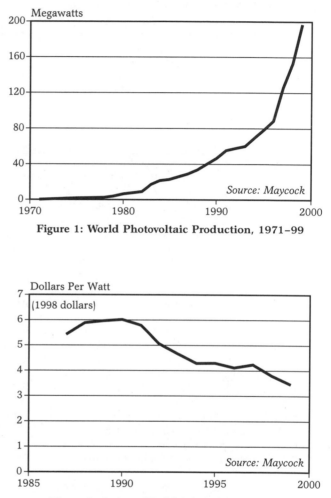

Figure 1: World Photovoltaic Production, 1971–99

Figure 2: Average World Wholesale Price for Photovoltaic Modules, 1988–99

Compact Fluorescents Light Up the Globe Michael Scholand

Global sales of energy-efficient compact fluorescent lamps (CFLs) grew a robust 11 percent in 1999, achieving record sales of 432 million units.[1] (See Figure 1.) CFLs, first commercialized in the early 1980s, are a miniaturized version of the familiar 4-foot fluorescent tubes.[2] But they last approximately 10 times longer than the incandescent light bulbs patented by Thomas Edison in 1880, and they use 75 percent less electricity while delivering the same amount of light.[3]

Since 1988, CFL sales have increased nearly 10-fold.[4] Because CFLs last for several years, there are an estimated 1.3 billion in use today running on 20,000 megawatts of electricity, instead of the 80,000 megawatts that would be needed to run the same number of incandescent lamps.[5] The electricity saved by CFLs each year is equivalent to that produced by 28 medium-sized coal-fired power plants.[6] In 1999, incandescents outsold CFLs by 25 to 1, but because CFLs last longer, they accounted for 33 percent of the lighting capacity sold.[7]

Electricity savings translate into avoided pollution. The 275 million CFLs being used in North America at the start of 2000, for instance, will avoid 3.5 million tons of carbon emissions and 69,000 tons of sulfur emissions during the year.[8] CFLs also reduce energy bills. In Denmark, consumers can pay 4 Euro ($4) for a high-quality CFL that, if lit four hours a day, pays back its additional cost in less than six months.[9] Looking at bulb replacement and electricity savings over the 10,000-hour life of the lamp, a CFL has a net present value of $49—12 times what it cost.[10]

Recognizing the benefits from CFLs, many governments are promoting them. China, for example, has just completed a three-year Green Lights program to expand the market for CFLs and improve their quality.[11] Between 1996 and 1999, China registered 347-percent growth in domestic sales while positioning itself as a global leader in manufacturing.[12] The Asian market for CFLs has more than doubled since 1995, and by 1999 it was twice the size of the North American market.[13] (See Figure 2.) The European Commission is planning its own Green Lights program involving all its member states, advocating efficient lighting in the residential and commercial sectors.[14]

The International Finance Corporation (IFC) has launched an Efficient Lighting Initiative (ELI) with support from the Global Environment Facility (GEF). ELI will work in seven countries—Argentina, the Czech Republic, Hungary, Latvia, Peru, the Philippines, and South Africa—promoting and expanding the market for CFLs and other efficient lighting technologies.[15] Russell Sturm of the IFC points out that "the rapid take-up of efficient lighting technology will not only reduce energy expenditure, but will allow the countries to meet their energy service needs more cost-effectively, freeing up scarce capital for other critical development needs, and reduce greenhouse gas emissions at a cost of less than $5 per ton of carbon."[16]

In support of the ELI program in South Africa, Eskom—the national electric utility—will match GEF funds, so an estimated $10 million will be spent there by 2003 promoting CFLs and other efficient lighting technologies.[17] Barry Bredenkamp, the project coordinator for Eskom, said "our efficient lighting program seeks to help all segments of the South African population. In the lower-income areas, we are combining the program with our on-going electrification drive, where we are literally helping households move from candles to compact fluorescents."[18]

CFL technology continues to improve, lowering costs and improving quality. The "ballast"—the bulky, complex electronics that enable the lamp to run—is the most expensive part of a CFL, accounting for 70 percent of the manufacturing cost.[19] Steve Johnson of Lawrence Berkeley Laboratory says that new circuit designs for CFL ballasts will make them smaller and less expensive, while also improving power quality and increasing light output. By 2002, he expects to see advanced designs that can fit in the metal screw-cap of a normal light bulb.[20]

WORLD SALES OF
COMPACT FLUORESCENT
BULBS, 1988–99

YEAR	SALES (million)
1988	45
1989	59
1990	83
1991	112
1992	138
1993	179
1994	206
1995	245
1996	288
1997	362
1998	387
1999 (prel)	432

SOURCE: Evan Mills, Lawrence Berkeley Laboratory, letter to Worldwatch, 3 February 1993; Nils Borg, IAEEL, e-mail, 14 January 2000.

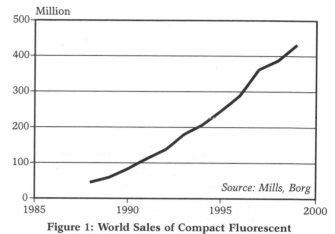

Figure 1: World Sales of Compact Fluorescent Bulbs, 1988–99

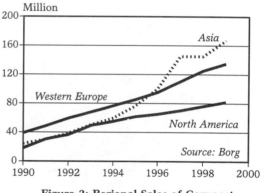

Figure 2: Regional Sales of Compact Fluorescent Bulbs, 1990–99

Atmospheric Trends

Global Temperature Drops

Seth Dunn

The average temperature of the atmosphere at Earth's surface fell to 14.35 degrees Celsius in 1999, according to land- and ocean-based measurements from the Goddard Institute for Space Studies of the National Aeronautics and Space Administration (NASA).[1] (See Figure 1.) Although this is 0.23 degrees below the record-setting 1998 average, it is the sixth highest of this dataset, which extends back to 1950, and the seventh highest in a land-based series dating to 1866.[2] (See Figure 2.)

Comparable instrumental measurements and "proxy" climate data corroborate and extend the NASA findings. The National Oceanic and Atmospheric Administration (NOAA) and the Hadley Centre of the U.K. Meteorological Office, with records to the 1880s and 1860s, found 1999 to be the fifth warmest year on record.[3] Tree-ring samples from researchers at the University of Massachusetts suggest that, at least in the northern hemisphere, the 1990s were the warmest decade of the millennium, and 1998 its warmest year.[4]

The drop in temperature last year was influenced by the onset of La Niña, a periodic cooling of tropical Pacific Ocean waters—whose flip side, El Niño, contributed to the 1998 record.[5] Ocean temperatures were the lowest since 1994, while land temperatures were second only to 1998.[6] Near-surface temperatures have been increasing by 0.2 degrees Celsius per decade since 1976.[7]

The high land mass measurements were driven mostly by continued warmth in Eurasia and North America.[8] Russia had one of its longest heat waves in a century, and parts of central and southern Europe experienced their warmest September in 100 years.[9] The United States had its second warmest year since 1900, with a monthly record set during November, also a record month globally.[10]

In the Southern Hemisphere, La Niña's presence led to cooler temperatures. Those in western and southern South America were near or below their long-term average.[11] Central and southern Africa—especially the Sahel region—were cooler than normal in the second half of the year.[12]

Scientists are exploring the relation between the overall warming trend and climate phenomena. NASA scientists observe that "the warmth of 1998 was too large and pervasive to be fully accounted for by the recent El Niño."[13] Noting the unique strength, duration, and pattern of the latest El Niño and La Niña, Michael McPhaden of NOAA suggests that human-caused global warming may be influencing the collective behavior of El Niño, La Niña, and an atmospheric pressure swing between the eastern and western Pacific.[14]

Another issue is the role of solar variability. Recent research suggests that the sun's magnetic field has doubled since 1900, and that this brightening may have contributed to temperature increases between 1910 and 1940.[15] But the effect of changes in solar irradiation on climate is likely to be small relative to the accelerated release of greenhouse gases from fossil fuel burning and other human activities, to which the increase since 1970 can be largely attributed.[16]

A January 2000 report from the U.S. National Research Council asserts that surface temperature measurements, taken daily for more than 100 years at hundreds of locations, are better indicators of long-term trends than comparatively short satellite records, which show little warming.[17] Any difference between the data, the report adds, does not invalidate the "undoubtedly real" warming of Earth's surface.[18]

The observed warming trend of about 0.6 degrees Celsius since the late 1800s is consistent with model predictions of the combined influences of greenhouse gases, solar activity, and sulfate aerosols.[19] Preliminary scenarios by the Intergovernmental Panel on Climate Change project a global mean temperature increase of 1.9–2.9 degrees Celsius between 1990 and 2100 (see Figure 3), with a corresponding rise in sea level of 46–58 centimeters.[20] This rise—whose range depends in part on future greenhouse gas emissions—is expected to cause uncertain but potentially major changes in precipitation patterns, water availability, and weather extremes.[21]

GLOBAL AVERAGE TEMPERATURE, 1950–99

YEAR	TEMPERATURE (degrees Celsius)
1950	13.84
1955	13.91
1960	13.96
1965	13.88
1970	14.02
1971	13.93
1972	14.01
1973	14.11
1974	13.92
1975	13.94
1976	13.81
1977	14.11
1978	14.04
1979	14.08
1980	14.18
1981	14.30
1982	14.09
1983	14.28
1984	14.13
1985	14.10
1986	14.16
1987	14.28
1988	14.32
1989	14.24
1990	14.40
1991	14.37
1992	14.20
1993	14.12
1994	14.22
1995	14.38
1996	14.32
1997	14.40
1998	14.58
1999 (prel)	14.35

SOURCES: Surface Air Temperature Analyses, Goddard Institute for Space Studies, New York, 11 January 2000.

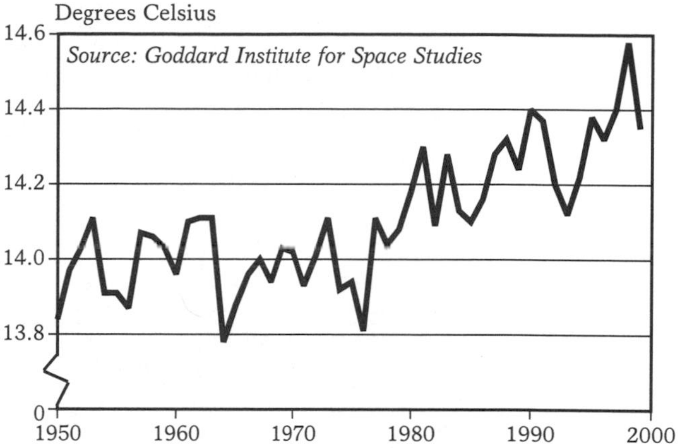

Figure 1: Average Temperature at Earth's Surface, 1950–99

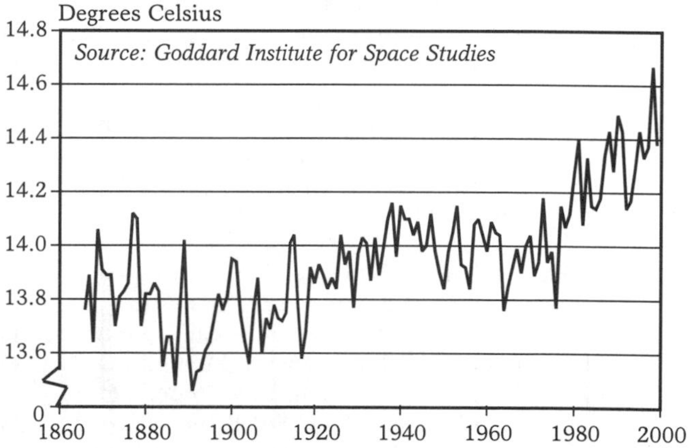

Figure 2: Average Temperature at Earth's Surface (Land-based Series), 1866–1999

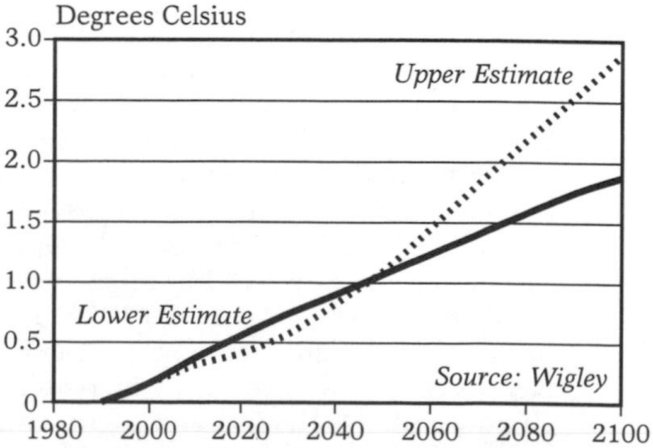

Figure 3: Mean Temperature Change, Preliminary IPCC Scenarios, 1990–2100

Carbon Emissions Fall Again   Seth Dunn

Emissions of carbon from fossil fuel combustion worldwide fell 0.2 percent to 6.31 million tons in 1999—the second consecutive annual decline in global carbon output.[1] (See Figure 1.) Carbon emissions growth averaged 0.6 percent during the 1990s, compared with 1.5 percent during the 1980s.[2] The "carbon intensity" of the global economy—emissions per unit of economic output—declined 38.8 percent between 1950 and 1999.[3] (See Figure 2.)

Developing economies saw a 34-percent increase in carbon output between 1990 and 1998.[4] Emissions in China and India rose 28 and 55 percent, respectively, during this period.[5] But while India's carbon intensity rose 2.9 percent between 1990 and 1996, China's fell 31.5 percent.[6]

Carbon emissions from industrial nations expanded 6.7 percent between 1990 and 1998.[7] Output in the United States grew 10.3 percent, while that in the European Union dropped 0.7 percent due to drops in Germany and the United Kingdom.[8] Japan, where emissions went up 5.6 percent between 1990 and 1998, boasts the world's lowest carbon intensity, which remained flat between 1990 and 1996.[9]

Carbon emissions fell 30.3 percent between 1990 and 1998 in former Eastern bloc nations, generally the most carbon-intensive economies.[10] Emissions in the region's largest emitter, Russia, shrank 29.3 percent between 1992 and 1998.[11] But Russian carbon intensity stands at more than three times that of China and over six times that of the European Union.[12]

The world's gradual decoupling of economic growth from carbon output is receiving attention as industrial and former Eastern bloc nations grapple with prospective commitments, under the 1997 Kyoto Protocol, to reduce emissions of greenhouse gases by 5 percent between 1990 and 2010.[13] Although carbon intensity declined by an average annual rate of 1.4 percent between 1970 and 1999, this pace is not sufficient to meet the Kyoto target—and would need to be increased severalfold to achieve the 70-percent cut many scientists believe necessary to avoid dangerous climate change.[14] Yet the transition to more information- and service-based economies, greater energy efficiency, and innovations such as hybrid-electric cars are improving prospects for an accelerated "decarbonization."[15]

During international climate negotiations in 1999, the European Union, Japan, and New Zealand announced plans to ratify the Kyoto Protocol by 2002.[16] Their ratifications, combined with that of the former Eastern bloc, would be enough to activate the protocol without the United States, where the Senate continues to oppose the pact.[17] Russian ratification is less certain, however, given its interest in trading its potentially large emissions surplus with other nations.[18]

Rules for emissions trading, which economists believe can lower the costs of reducing carbon emissions, are to be fleshed out at negotiations in November 2000.[19] The future commitments of developing countries are likely to be an important unofficial subject of debate. At the talks in 1999, Argentina became the first developing nation to set greenhouse gas emissions limits voluntarily.[20]

The cumulative release of carbon pushed atmospheric carbon dioxide (CO_2) concentrations up to 368.4 parts per million volume (ppmv)—31.6 percent above the pre-industrial level of 280 ppmv.[21] (See Figure 3.) The average annual increase of CO_2 levels during the 1990s was 1.6 ppmv per year, with the 1998 rise of 2.8 ppmv a record for the last 40 years.[22] Though emissions from human activities are primarily responsible, oceanic variability makes CO_2 growth rates higher than normal during El Niño years and lower during La Niña years.[23]

A 1999 *Nature* article based on new data from the ice core in Vostok, Antarctica—the deepest ice core ever drilled—suggests that current CO_2 levels have been "unprecedented" during the past 420,000 years.[24] The Vostok data also confirm that past increases in CO_2 concentrations have contributed to major global warming transitions at Earth's surface.[25]

WORLD CARBON EMISSIONS FROM
FOSSIL FUEL BURNING, 1950–99,
AND ATMOSPHERIC CONCENTRATIONS
OF CARBON DIOXIDE, 1960–99

YEAR	EMISSIONS (mill. tons of carbon)	CARBON DIOXIDE (parts per mill.)
1950	1,612	n.a.
1955	2,013	n.a.
1960	2,535	316.7
1965	3,087	319.9
1970	3,998	325.5
1971	4,143	326.2
1972	4,306	327.3
1973	4,538	329.5
1974	4,545	330.1
1975	4,518	331.0
1976	4,777	332.0
1977	4,910	333.7
1978	4,950	335.3
1979	5,229	336.7
1980	5,156	338.5
1981	4,984	339.8
1982	4,947	341.0
1983	4,933	342.6
1984	5,098	344.2
1985	5,271	345.7
1986	5,453	347.0
1987	5,575	348.7
1988	5,789	351.3
1989	5,892	352.7
1990	5,946	354.0
1991	6,021	355.5
1992	5,928	356.3
1993	5,896	357.0
1994	6,034	358.8
1995	6,212	360.9
1996	6,316	362.6
1997	6,349	363.9
1998	6,318	366.6
1999 (prel)	6,307	368.4

SOURCES: Worldwatch estimates based on ORNL, BP, DOE, EC, Eurogas, PlanEcon, IMF, and LBL.

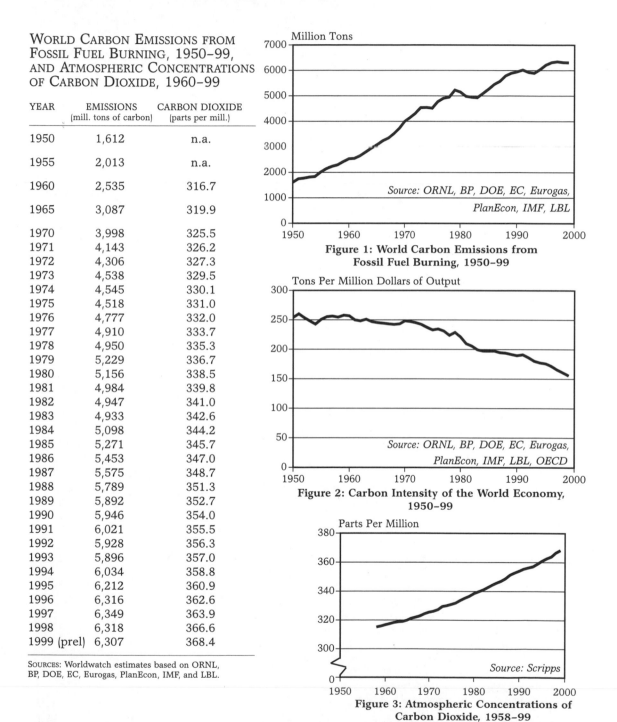

Million Tons

Source: ORNL, BP, DOE, EC, Eurogas, PlanEcon, IMF, LBL

Figure 1: World Carbon Emissions from Fossil Fuel Burning, 1950–99

Tons Per Million Dollars of Output

Source: ORNL, BP, DOE, EC, Eurogas, PlanEcon, IMF, LBL, OECD

Figure 2: Carbon Intensity of the World Economy, 1950–99

Parts Per Million

Source: Scripps

Figure 3: Atmospheric Concentrations of Carbon Dioxide, 1958–99

Economic Trends

Economic Growth Speeds Up
David Malin Roodman

Gross world product (GWP) rose 3.0 percent in 1999, to $40.5 trillion (in 1998 dollars).[1] (See Figure 1.) This growth rate is lower than the 4-percent pace of the mid-1990s, but a bit higher than 1998's 2.5-percent increase.[2]

Almost all of the acceleration occurred in Asia, which began recovering from the "Asian flu" of 1997–98. In South Korea, gross domestic product (GDP) rose 6.5 percent in 1999, after having fallen 5.8 percent the year before.[3] In Singapore, meanwhile, growth went from 0.3 percent to 4.5 percent, and in Hong Kong it went from –5.1 percent to +1.2 percent.[4] Japan, whose economy is tied to the East Asian "tigers" with which it trades and competes, also saw a turnaround: its GDP expanded slightly in 1999, having shrunk 2.8 percent in 1998.[5] Because Japan's GDP is so large ($3.1 trillion), this seemingly less dramatic shift accounted for more than half of the global growth pick-up in 1999.[6]

Russia, which had also been struck by crisis, recovered somewhat in 1999—from –4.6 percent "growth" to 0.0 percent.[7] Latin America also saw essentially no growth in 1999—0.1 percent. The year began with Brazil teetering on the brink of financial panic.[8]

Other regions experienced modest slowdowns. In Africa, growth declined from 3.4 percent to 3.1 percent.[9] It fell from 2.2 percent to 1.0 percent in Central and Eastern Europe (excluding Russia), from 2.7 percent to 2.0 percent in the European Union, and from 3.9 percent to 3.7 percent in the United States.[10]

The experiences of nations recently in economic crisis show how growth-rate changes indicate important developments. In Indonesia, GDP shrank an extraordinary 13.7 percent in 1998, the same year the share of people in poverty reportedly climbed from 11 percent to 18 percent.[11] Similarly, growth slowdowns in western industrial countries have historically coincided with mass layoffs and rising unemployment.[12]

Still, critics of the GDP statistic argue that obsession with that number leads policymakers to distort government action to favor what GDP counts and to neglect or undermine what it does not—to encourage commerce, say, at the expense of the environment.[13] Most trends for which GDP obsession has received blame—from ecological harm to widening rich-poor gaps—actually predate it. But GDP has indeed become a modern totem: political leaders everywhere covet growth for its reputed ability to heal poverty, political unrest, even air pollution.[14]

One criticism of GDP is that it is indifferent as to whether a dollar of income goes to a billionaire or a pauper—either way, total income is the same. For example, GDP per person rose 71.5 percent in the United States between 1967 and 1997.[15] But the richest 20 percent of households benefited the most: they earned 11 times as much as the poorest 20 percent in 1967 and 13 times as much in 1997.[16] A GDP alternative maintained by Redefining Progress in San Francisco, the Genuine Progress Indicator, was 30 percent lower in 1997 than it would have been had inequality not risen.[17]

The GDP-hidden differences between rich and poor are equally extreme across nations. The global economy pumped out nearly $6,800 of goods and services per person in 1999—but 45 percent of the income went to the 12 percent of people living in western industrial nations.[18] There, GDP per capita averaged $25,000—compared with $6,400 in Latin America, $5,100 in the former Eastern bloc, $4,400 in Asia (including Japan), and $1,600 in Africa. (See Figure 2.)

GDP accounting also ignores or undercounts natural "goods and services," from fertile soil and fresh water to climate regulation—treating them as cheap or free. Yet a group of scientists has estimated that they are worth some $36 trillion a year, on a par with GWP.[19]

Also left out is the housework and child care done, mostly by women, outside the cash economy—even though these services are counted when they are provided commercially. The U.N. Development Programme estimated that the unpaid work of women was worth $12 trillion worldwide in 1995.[20] All in all, GWP misses much of the true global economy.

Economic Growth Speeds Up

GROSS WORLD PRODUCT, 1950–99

YEAR	TOTAL (trill. 1998 dollars)	PER PERSON (1998 dollars)
1950	6.3	2,525
1955	8.0	2,911
1960	9.9	3,262
1965	12.6	3,800
1970	16.1	4,393
1971	16.8	4,475
1972	17.5	4,594
1973	18.7	4,803
1974	19.1	4,819
1975	19.4	4,791
1976	20.4	4,943
1977	21.2	5,069
1978	22.1	5,185
1979	22.8	5,265
1980	23.3	5,278
1981	23.8	5,292
1982	24.0	5,262
1983	24.7	5,319
1984	25.8	5,463
1985	26.7	5,551
1986	27.6	5,641
1987	28.6	5,743
1988	29.9	5,887
1989	30.7	5,960
1990	31.4	5,965
1991	31.5	5,893
1992	31.9	5,854
1993	32.7	5,923
1994	34.0	6,073
1995	35.3	6,215
1996	36.8	6,394
1997	38.4	6,572
1998	39.3	6,647
1999 (prel)	40.5	6,757

SOURCES: Worldwatch update of Angus Maddison, *Monitoring the World Economy 1820–1992* (Paris: OECD, 1995); updates from IMF, *World Economic Outlook* tables.

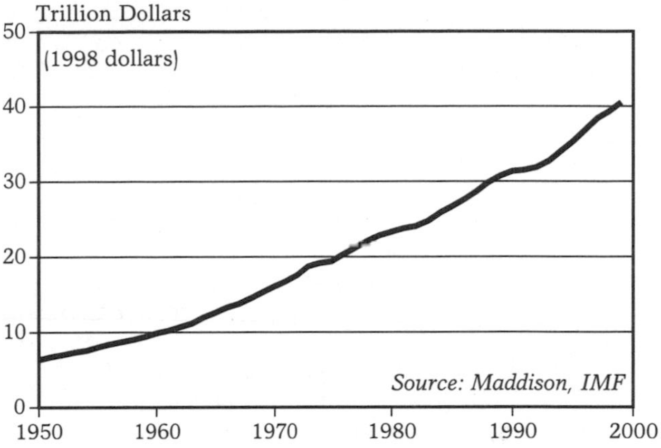

Figure 1: Gross World Product, 1950–99

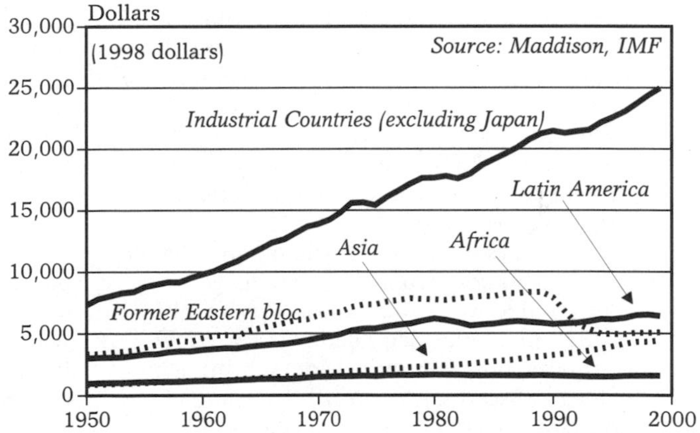

Figure 2: Gross Economic Product Per Capita by Region, 1950–99

Developing-Country Debt Increases　　　Sarah Porter

In 1998, the latest year for which figures are available, the total external debt burden of developing countries increased to an estimated $2.5 trillion, up more than 5 percent from 1997.[1] (See Figure 1.) East Asia and Latin America account for more than 58 percent of the developing-world debt burden.[2] Eastern Europe and the former Soviet republics hold 18 percent of the total, while sub-Saharan Africa carries 9 percent.[3] Low- and middle-income countries in South Asia, the Middle East, and North Africa carry the remaining 15 percent.[4]

Long-term debt accounts for 79 percent of all external debt, with the rest being short-term debt and the use of credit from the International Monetary Fund (IMF).[5] Of long-term debt, 57 percent was owed to commercial banks and other creditors, 27 percent to other governments, and 16 percent to multilateral agencies such as the World Bank and the regional development banks.[6] But these figures vary greatly across regions, as private lenders have been reluctant to make loans to countries that are perceived as risky and low-performing. In Latin America and the Caribbean, 72 percent of debt was owed to private creditors, while in sub-Saharan Africa the figure was only 24 percent.[7]

Annual debt service payments in 1998 fell nearly 4 percent from the previous year, though they were still almost 50 percent higher than in 1990.[8] (See Figure 2.) For many countries, however, debt servicing still consumes a disproportionately large share of government revenues compared with funds devoted to basic social services. In Zambia, for example, only 10 percent of government spending goes for health, education, and other basic social services, while 30 percent is applied to paying off foreign debt.[9] And in many instances, countries have ceased to make payments on their outstanding loans. Jubilee 2000, an international campaign calling for debt cancellation, estimates that 55 percent of all loans to poor countries are not being serviced at all.[10]

One indicator of a country's ability to manage debt successfully is the ratio of total external debt to annual export earnings, since debt servicing draws on the foreign exchange earned from exports.[11] For developing countries as a whole, this ratio has gradually fallen since the late 1980s, reaching 146 percent in 1998.[12] (See Figure 3.) Yet for the 40 nations designated as Highly Indebted Poor Countries (HIPC)—31 of which are in Africa—this ratio is at least two and a half times higher.[13]

In recent years a growing movement of churches and civil society organizations, many united under the Jubilee 2000 umbrella, have pressed for complete cancellation of debts owed by the world's poorest countries. They argue that previous measures for debt relief and rescheduling have failed to reduce the overall debt burden, and that excessive debt continues to retard or even reverse economic growth in many countries.[14]

In creditor countries, political support for debt relief has grown over the last few years. By late 1999, the United Kingdom, the United States, and Canada had pledged to cancel 100 percent of bilateral debts owed by poor countries.[15] Earlier in 1999, at a meeting of the Group of Seven in Cologne, Germany, the leading industrial countries agreed to provide $100 billion in debt relief.[16] Under this package, creditors would write off about two thirds of the official debt owed by the world's poorest countries, both by canceling bilateral debts as well as by reforming the IMF/World Bank HIPC Initiative.[17] Launched in 1996, the HIPC Initiative aims to reduce debts to "sustainable" levels. It requires countries to meet economic reform targets, often by implementing structural adjustment policies that call for cuts in government spending, privatization, and trade liberalization.[18] Some critics charge that these offers are inadequate, however, as they still require recipients to implement a range of structural adjustment policies that can exacerbate poverty.[19]

EXTERNAL DEBT AND DEBT SERVICE OF ALL DEVELOPING COUNTRIES, 1971–98

YEAR	DEBT	DEBT SERVICE
	(bill. 1998 dollars)	
1971	295	34
1972	332	38
1973	377	49
1974	432	55
1975	524	58
1976	601	65
1977	751	80
1978	879	109
1979	997	136
1980	1,138	174
1981	1,199	186
1982	1,298	195
1983	1,352	178
1984	1,364	184
1985	1,494	195
1986	1,595	201
1987	1,753	208
1988	1,695	216
1989	1,714	202
1990	1,773	198
1991	1,808	188
1992	1,843	188
1993	1,976	193
1994	2,138	214
1995	2,267	254
1996	2,304	288
1997	2,340	308
1998	2,465	296

SOURCE: World Bank, *Global Development Finance 1999*, electronic database, Washington, DC, 1999.

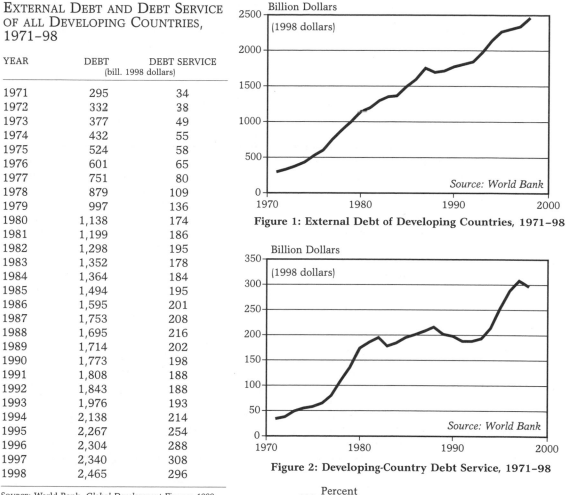

Figure 1: External Debt of Developing Countries, 1971–98

Figure 2: Developing-Country Debt Service, 1971–98

Figure 3: Developing-Country Debt as Share of Exports, 1971–98

World Trade Stable in Value

David Malin Roodman

The value of world exports rose slightly between 1998 and 1999 to $6.8 trillion (in 1998 dollars).[1] (See Figure 1.) Exports of goods, ranging from copper to cars, totaled $5.4 trillion in 1999, while exports of services such as air travel and banking were $1.3 trillion.[2]

Between 1980 and 1995, exports leapt from $4.4 trillion to $6.6 trillion.[3] But after 1995, goods exports, which account for most of the total, rose in volume but fell in price, especially following currency crashes in East Asia. An International Monetary Fund index that tracks dollar prices of traded goods fell 12 percent between 1995 and 1998.[4] The value of 1998's actual exports would have been 18 percent higher than that for 1995 had prices instead risen gradually at the U.S. inflation rate.[5]

In 1950, the United States generated a sixth of world goods exports.[6] (See Figure 2.) But its share shrank during the next 30 years as Japan, West Germany, and others emerged from the ashes of World War II.[7] Saudi Arabia's share of goods exports shot to 4.4 percent in 1974 after the first oil shock, and to 6.2 percent in 1981 after the second—but then collapsed along with oil prices by the mid-1980s.[8] China's share rose steadily starting in the late 1970s as market-liberalizing reforms took hold, reaching 3.5 percent in 1999.[9]

Although exports of tangible goods dominate global trade tallies, services also play an important role. Most exported services help move people and goods across borders. In 1998, for instance, customers in one country paid freight companies in another country $134 billion to ship cargo.[10] Travelers purchased $84 billion of transportation from foreign carriers, and spent $425 billion on everything from hotel rooms to souvenirs once they were on foreign soil.[11]

Some of the fastest export growth is occurring in relatively minor service categories. U.S. exports of construction services, for example, more than doubled between 1991 and 1998.[12] Also doubling were U.S. exports of financial services such as banking, as well as licensing fees and royalties for software, movies, and other intellectual property.[13]

Between 1995 and 1999, exports of all goods and services fell from 18.4 percent of gross world product to 16.8 percent, because the dollar value of economic output rose faster than that of trade.[14] (See Figure 3.) But this modest drop may prove only a pause in a long upward trend if exports in the less traditional categories such as banking continue growing rapidly.

How much further the global economy will integrate is far from certain. The steady increase in trade since 1950 arose as much from international political will as from economic forces. The years 1947–94 saw eight rounds of negotiations on the General Agreement on Tariffs and Trade to reduce trade barriers from high prewar levels.[15] The last, "Uruguay" Round took six years and ended in 1994. The resulting agreement created the World Trade Organization (WTO) and took major steps on international investment, intellectual property rights, trade in services, and other areas important to industrial nations.[16] But the treaty was less aggressive on issues of importance to developing countries, such as reducing rich countries' protections for their agriculture and textile industries.[17]

The Uruguay Round also strengthened the system for resolving trade disputes.[18] Once a WTO panel makes a decision, the winner of the case may hit the loser with countervailing trade measures—as when the United States slapped steep duties on Roquefort cheese and other foods from the European Union (EU) after a panel ruled against the EU import ban on hormone-fed beef.[19]

Yet the WTO has ruled against similar measures that are meant to protect international norms not narrowly relating to trade—including a U.S. import ban against tuna caught with dolphin-ensnaring nets.[20] In this sense, the WTO has effectively placed trade ahead of the environment, human rights, and all other important international concerns. The public perception in rich countries that the WTO distorts public priorities, along with widespread doubt among poorer countries about whether the treaty serves their interests, in no small part stymied the launch of a new negotiation round in 1999, in Seattle.

WORLD EXPORTS OF GOODS, 1950–99, AND GOODS AND SERVICES, 1980–99

YEAR	GOODS	GOODS AND SERVICES
	(trill. 1998 dollars)	
1950	0.4	
1955	0.5	
1960	0.6	
1965	0.8	
1970	1.1	
1971	1.2	
1972	1.3	
1973	1.8	
1974	2.4	
1975	2.3	
1976	2.4	
1977	2.6	
1978	2.8	
1979	3.3	
1980	3.6	4.4
1981	3.3	4.0
1982	2.9	3.5
1983	2.7	3.3
1984	2.7	3.3
1985	2.7	3.3
1986	2.9	3.5
1987	3.3	4.0
1988	3.6	4.4
1989	3.8	4.6
1990	4.1	5.1
1991	4.0	5.0
1992	4.2	5.2
1993	4.1	5.2
1994	4.6	5.7
1995	5.4	6.6
1996	5.5	6.8
1997	5.6	6.9
1998	5.4	6.7
1999 (prel)	5.5	6.8

SOURCES: IMF, *International Financial Statistics*, electronic database, February 2000; Barbara d'Andrea-Adrian, WTO, e-mail to author, 16 February 2000.

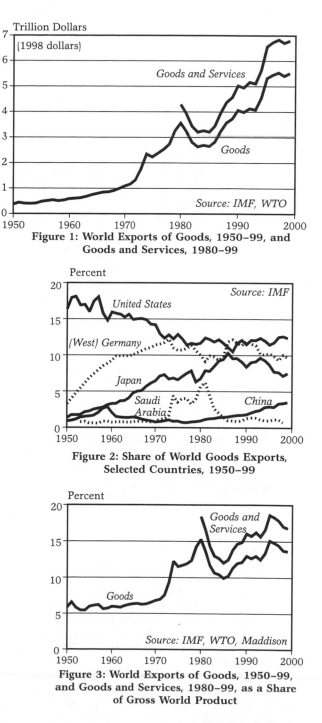

Figure 1: World Exports of Goods, 1950–99, and Goods and Services, 1980–99

Figure 2: Share of World Goods Exports, Selected Countries, 1950–99

Figure 3: World Exports of Goods, 1950–99, and Goods and Services, 1980–99, as a Share of Gross World Product

Weather Damages Drop

Seth Dunn

Storms, floods, and other weather-related disasters caused $67 billion in economic losses in 1999, based on estimates from the Munich Reinsurance Company.[1] This was the second-costliest year on record, after 1998.[2] (See Figure 1.) Total weather-linked losses in the 1990s topped $430 billion—more than five times the figure for the 1980s, in 1998 dollars.[3] More than 52,000 deaths could be attributed to these events in 1999—compared with a 1998 estimate of 41,000 weather-related mortalities.[4]

Insured weather-related losses, totaling just under $20 billion in 1999—the second-highest total after 1992—reached $112 billion in the 1990s, a fourfold increase from the previous decade.[5] (See Figure 2.)

The most costly weather event overall in 1999, in both economic and human terms, was a devastating December rainstorm over northern Venezuela—one of Latin America's worst natural disasters of the last century.[6] The subsequent floods and landslides, combined with uncontrolled logging and human settlement in the mountains, caused close to $15 billion in economic losses and more than 30,000 deaths.[7]

The year's most costly insured event was December's series of winter storms that swept through Central and Western Europe—especially France, Germany, Spain, and Switzerland—resulting in the continent's most expensive disaster in a decade and leading to $9.6 billion in damages, $5.1 billion of which were insured losses.[8] At least 130 people died from the storms, whose quick succession was deemed by one meteorologist a "once in a hundred year" event.[9]

In October, a cyclone ravaged Orissa, India, causing $2.5 billion in losses and taking 15,000 lives.[10] The Orissa cyclone, a "super-cyclone" that reached winds of up to 300 kilometers per hour, was one of the strongest to hit the Indian subcontinent in a century.[11] The storm affected more than 20 million people, including more than 100,000 residents of the provincial capital of Bhubanseshwar who lost their homes.[12]

Limited preparation for the cyclone, and difficulties in providing aid to refugees, drew attention to India's lack of a disaster-management policy.[13] But richer nations are also deficient in this area. The U.S. National Academy of Sciences, documenting the dramatic rise in natural disaster costs during the past quarter-century in the United States, notes that short-term preventive strategies often raise the risk of future catastrophes.[14] Levees built along the Mississippi River, for example, led to intensive building in floodplains; thousands of homes were subsequently destroyed by a flood in 1993.[15] Preparations for Hurricane Floyd, which caused $4.1 billion in damages along the Bahamas and U.S. East Coast in 1999, led to the largest evacuation in U.S. history. But thousands were sent into flood-prone regions, while others were stuck in traffic jams; an incorrect storm prediction, meanwhile, left unsuspecting communities vulnerable as the hurricane traveled further inland than forecast.[16]

The human factor in weather-related losses causes considerable debate. Research from the National Oceanic and Atmospheric Administration suggests that increases in U.S. economic losses and fatalities from weather events over the last 25 years are primarily due to societal changes such as population growth and greater property exposure in coastal areas.[17] A study prepared for the Canadian government, however, notes that climate-related losses are increasing many times more rapidly than earthquake disaster losses, and that differences in economic development between earthquake- and weather-prone regions are unlikely to account for this gap.[18]

Whatever their separate roles, environmental and social problems can in combination create major weather catastrophes. The International Federation of Red Cross and Red Crescent Societies predicts "super-disasters" in the coming decade, arising from the impact of climate change–driven catastrophes on people made increasingly vulnerable by poverty, urbanization, ecological degradation, and other changes.[19] Half of the world's population lives in coastal regions—putting billions at risk from rising sea levels and more frequent and severe floods and cyclones.[20]

ECONOMIC LOSSES FROM WEATHER-RELATED NATURAL DISASTERS WORLDWIDE, TOTAL AND INSURED, 1980–99

YEAR	TOTAL LOSSES (bill. 1998 dollars)
1980	2.8
1981	13.3
1982	3.4
1983	9.5
1984	3.4
1985	7.2
1986	9.4
1987	13.0
1988	4.2
1989	12.2
1990	18.0
1991	31.2
1992	40.5
1993	24.4
1994	24.1
1995	40.3
1996	61.7
1997	30.3
1998	92.9
1999 (prel)	67.1

YEAR	INSURED LOSSES (bill. 1998 dollars)
1980	0.1
1981	0.6
1982	1.5
1983	4.5
1984	1.5
1985	2.9
1986	0.3
1987	5.8
1988	1.0
1989	5.6
1990	12.0
1991	9.3
1992	25.3
1993	5.8
1994	1.9
1995	9.4
1996	9.3
1997	4.5
1998	15.1
1999 (prel)	19.7

SOURCE: Munich Re database and e-mail, 3 February 2000.

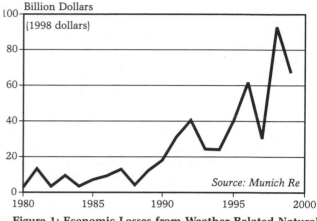
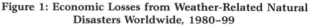

Figure 1: Economic Losses from Weather-Related Natural Disasters Worldwide, 1980–99

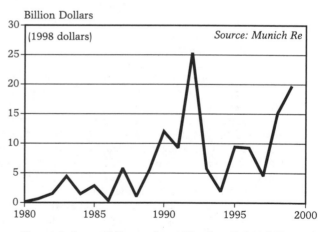

Figure 2: Insured Losses from Weather-Related Natural Disasters Worldwide, 1980–99

Paper Piles Up Ashley T. Mattoon

The U.N. Food and Agriculture Organization (FAO) reports that world production of paper and paperboard rose from approximately 289 million tons in 1997 to 294 million tons in 1998, the latest year for which data are available.[1] (See Figure 1.) Production has increased more than sixfold since 1950, and per capita consumption has jumped from about 18 kilograms to 50 kilograms.[2] By the end of this decade, production is expected to reach 394 million tons, a 34-percent increase over 1998's level.[3]

The world's leading paper producers are the United States, Japan, and China—accounting for 29, 10, and 9 percent, respectively, of total world production.[4] These three are also the leading consumers, with each using about the same proportion as they produce.[5]

Paper use is closely correlated with income levels, and most of the world's paper is consumed in industrial countries. With close to 22 percent of the world's population, these nations account for more than 71 percent of paper use.[6] On a per capita basis, the gap in paper use is even wider. (See Figure 2.) In 1997, annual per capita paper consumption in the United States was 335 kilograms, and for industrial countries overall, the average was 162 kilograms.[7] In contrast, the global average in 1997 was 51 kilograms per person a year, and for developing nations it was 18 kilograms.[8] (One kilogram is roughly equivalent to 225 sheets of office paper, or two copies of a daily *New York Times*.)

Since the 1960s, the volume of trade in pulp and paper has increased more than fivefold.[9] Today, 29 percent of the paper that is produced in the world is traded internationally and paper products represent close to 45 percent of the total value of world forest products exports.[10]

Paper is used for hundreds of different purposes, and it is most often viewed as an ephemeral, disposable material. Only about 10 percent of the paper that is produced goes to making long-lasting products like books.[11] The other 90 percent is used just once and discarded. In 1997, almost half of the paper used was for packaging.[12] Printing and writing papers accounted for 30 percent of paper use, newsprint another 12 percent, and sanitary and household papers made up 6 percent.[13] While the use of all types of papers has increased over time, consumption of printing and writing paper in recent years has grown faster than grades such as packaging paper and newsprint. Since 1980, global paper consumption has jumped by 75 percent while printing and writing paper consumption rose by 112 percent.[14]

Today, virgin wood fiber makes up 55 percent of the total fiber supply for paper. Recycled paper contributes 38 percent, and nonwood fibers such as wheat straw and hemp account for the remaining 7 percent.[15]

The virgin wood that is used to make paper accounts for almost one fifth of the world's total wood harvest.[16] Of the wood harvested for "industrial" uses (everything but fuel), fully 42 percent becomes paper.[17] This proportion is expected to grow in the coming years since the world's appetite for paper is expanding faster than for other major wood products.[18] By 2050, it is expected that pulp and paper manufacture will account for over half of the world's "industrial" wood demand.[19]

About 54 percent of the wood that is used for making paper comes from second-growth forests.[20] Roughly 17 percent comes from original old-growth forests—primarily those in boreal regions of Canada and Russia.[21] Pulpwood plantations make up the remaining 29 percent, though this share is growing quickly.[22]

In addition to the demand on forests, paper production uses large amounts of energy, water, and chemicals, and generates vast amounts of air and water pollution and solid waste.[23] Worldwide, pulp and paper is the fifth largest industrial energy consumer, and accounts for about 4 percent of the world's total energy use.[24] In the United States, the pulp and paper industry is ranked third in the release of toxic chemicals to the environment—behind the chemicals and primary metals sectors.[25] In some industrial countries, paper makes up close to 40 percent of the total municipal solid waste burden.[26]

WORLD PAPER AND PAPERBOARD PRODUCTION, 1961–98

YEAR	PRODUCTION (mill. tons)	PER PERSON (kilograms)
1961	77	25
1962	81	26
1963	86	27
1964	92	28
1965	98	29
1966	105	31
1967	106	30
1968	114	32
1969	123	34
1970	126	34
1971	128	34
1972	138	36
1973	148	38
1974	150	37
1975	131	32
1976	147	35
1977	152	36
1978	160	37
1979	169	39
1980	170	38
1981	171	38
1982	167	36
1983	177	38
1984	190	40
1985	193	40
1986	203	41
1987	215	43
1988	228	45
1989	233	45
1990	240	46
1991	243	45
1992	245	45
1993	252	46
1994	268	48
1995	282	50
1996	282	49
1997	289	50
1998	294	50

SOURCE: FAO, *FAOSTAT Statistics Database*, Rome.

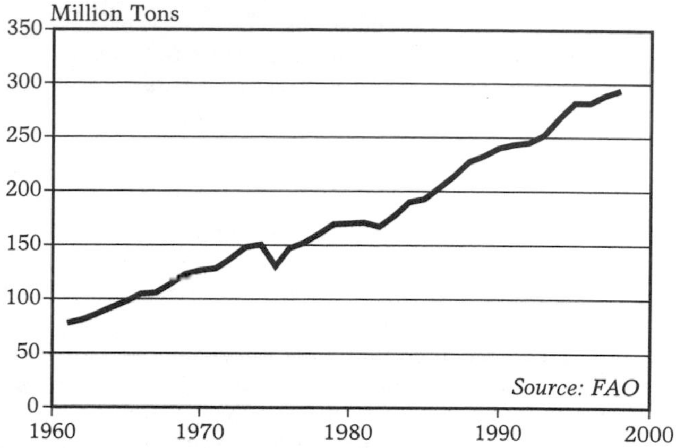

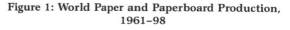

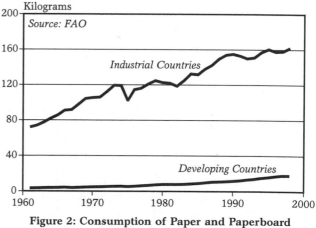

Figure 1: World Paper and Paperboard Production, 1961–98

Figure 2: Consumption of Paper and Paperboard Per Person, Industrial and Developing Countries, 1961–98

Gold Loses Its Luster

Payal Sampat

As gold prices dipped to their lowest levels since 1973, mines around the world put the brakes on production of this precious metal.[1] Global gold output shrunk by 5 percent in 1999, following two decades of steady expansion.[2] (See Figure 1.) Several unprofitable mines were abandoned in the United States, South Africa, Australia, and other key gold-mining nations, leaving local communities to deal with job losses and the toxic legacy of centuries of gold extraction.[3]

The 2,330 tons that were mined in 1999 are dwarfed by above-ground supplies.[4] Of the 137,000 tons of gold stocks held above-ground, nearly one quarter lies in national bank vaults—which alone is equal to 14 years of current mine output.[5] Another 18 percent is held by private investors, and almost half of the gold ever mined has been made into jewelry.[6] Today, almost 80 percent of gold is used for jewelry, sold mainly in Asia and the Middle East.[7] In addition to newly mined gold, 624 tons of scrap and 441 tons of officially held gold were added to the supply stream in 1999.[8]

In recent years, central banks and other major investors have dumped their gold stocks onto the world market. This has driven down prices—which hovered at around $275 an ounce in 1999—in turn spurring further disinvestment.[9] (See Figure 2.) Since January 1997, when the Dutch announced the sale of a quarter of their gold reserves, the world's national banks have jettisoned a total of 1,284 tons of the metal.[10] (See Figure 3.) Investors' confidence was further jolted when leading gold producer Australia sold two thirds of its reserves.[11] In May 1999, the Bank of England decided to discard 415 tons—more than half its reserves—into the market, following the Swiss National Bank's proposal to sell 1,300 tons over five years.[12]

Since ancient times, gold has been considered equivalent to money, and for 150 years, major world currencies were tied to the price of gold—a system known as the "gold standard."[13] Recent central bank sales reflect a growing mistrust in the investment value of gold, which *The Economist* has labeled "the

spent fuel of an obsolete monetary system."[14] Indeed, some argue that the single largest investor, the U.S. government, which holds 8,170 tons, has lost billions of dollars by keeping its reserves after the gold standard ended in 1971.[15] Many analysts argue that gold is now making a transition from a form of currency to a commodity like tin or iron—which the investment firm Lehman Brothers calls "reverse alchemy," the transformation of gold into a base metal.[16]

As prices tumbled, several uneconomical operations were closed in top producing countries, including the Timbarra mine in Australia and Leadville in the United States.[17] In contrast, production was stepped up in developing countries such as Indonesia and Peru, where rich veins, cheap labor, and weak environmental laws keep mining costs down.[18] The price slide also cramped global metals exploration spending—half of which was for gold—which fell by 48 percent between 1997 and 1999.[19]

Mines in South Africa, the leading gold producer, are among the most costly to operate, burrowing deep underground and using a lot of labor.[20] As ore grades there have declined, operating costs have often exceeded the market price of gold.[21] In 1999, South African mines laid off 100,000 workers—a third of the total—as many operations mechanized or closed down.[22]

Gold mining leaves a toxic trail that threatens human and ecological health. In one of the world's most hazardous professions, each ton of gold mined in South Africa on average causes one worker death and 11 serious injuries.[23] An ounce of marketable gold leaves in its wake nine tons of chemical-laced waste, polluted streams, and scarred landscapes.[24] In January 2000, a giant cyanide spill from a Romanian gold mine damaged 400 miles of river, poisoning water supplies and killing fish; this came close on the heels of similar catastrophes in Guyana, Kyrgyzstan, and Colorado in the United States.[25] As of 1993, the cleanup bill for the 557,000 abandoned hardrock U.S. mining sites was estimated at between $32 billion and $72 billion.[26]

GLOBAL GOLD PRODUCTION, 1950–99

YEAR	AMOUNT (tons)
1950	1,016
1955	1,118
1960	1,160
1965	1,438
1970	1,478
1971	1,446
1972	1,395
1973	1,347
1974	1,248
1975	1,197
1976	1,214
1977	1,210
1978	1,212
1979	1,207
1980	1,209
1981	1,283
1982	1,341
1983	1,405
1984	1,460
1985	1,532
1986	1,602
1987	1,658
1988	1,848
1989	1,971
1990	2,050
1991	2,110
1992	2,248
1993	2,290
1994	2,250
1995	2,210
1996	2,260
1997	2,400
1998	2,460
1999 (prel)	2,330

SOURCE: U.S. Geological Survey, January and February 2000.

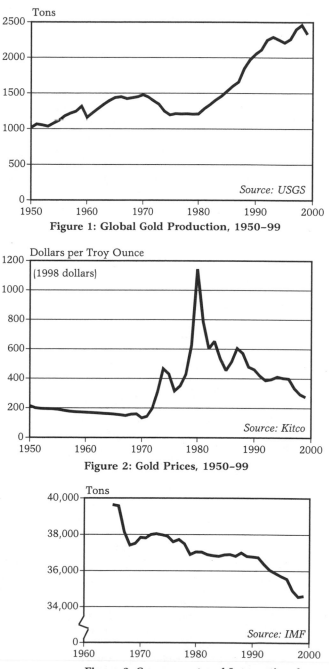

Figure 1: Global Gold Production, 1950–99

Source: USGS

Figure 2: Gold Prices, 1950–99

Source: Kitco

Figure 3: Government and International Institution Gold Holdings, 1965–99

Source: IMF

Tourism Growth Rebounds Lisa Mastny

International tourism increased 3 percent in 1999, reaching a record high of 657 million arrivals, according to estimates by the World Tourism Organization.[1] (See Figure 1.) This marked a strong rebound following two years of slowed growth during the global financial crisis. Receipts from international tourism (excluding transport) also rose again in 1999—to $449 billion (in 1998 dollars).[2] (See Figure 2.)

Tourism is a booming global pastime. International tourist arrivals have increased nearly 27-fold since 1950, at an average annual rate of 7 percent.[3] Europe is still the top destination, with nearly 60 percent of international arrivals in 1999.[4] (See Figure 3.) France was the most visited country that year, followed by Spain, the United States, Italy, and China.[5]

The United States earned the highest tourism receipts worldwide in 1999, and dominates tourism in the Americas.[6] But arrivals grew only 1 percent in 1999, while those to Canada rose 4 percent, bolstered by the weak Canadian dollar.[7] Latin America and the Caribbean have seen much faster growth: visits to Cuba alone have more than doubled since 1995.[8]

Asian tourism is recovering rapidly from the recent economic crisis.[9] Asia now attracts 14 percent of all international arrivals, compared with less than 1 percent in 1950.[10] Arrivals are expected to double within the next decade, and by 2020 the region could attract a quarter of all tourism traffic.[11] In the past decade alone, China has risen from twelfth to fifth place on the list of most visited nations.[12] It is predicted to be the top destination by 2020.[13]

Globally, international arrivals are expected to double by 2020—and governments, tour companies, and hotels are scurrying to meet demand.[14] Between 1980 and 1997, the number of hotel beds worldwide jumped by 80 percent, to more than 29 million.[15] Asia recorded the fastest growth, though Europe still boasts the most beds overall.[16] In 1999, total capital investments in travel and tourism neared $733 billion.[17]

Overall, tourism-related spending accounted for some $4.5 trillion of global economic activity in 1999—12 percent of the gross world product, according to the World Travel and Tourism Council.[18] Tourism also helped create an estimated 192 million jobs, or 8 percent of new jobs worldwide.[19] Every hotel room added is thought to create one to two posts, so a 300-room hotel can bring as many as 600 new jobs.[20]

Developing countries have much to gain from the tourism boom.[21] Central America and the Middle East were the two fastest-growing tourist regions in 1999, attracting some 23 percent and 18 percent more arrivals than in 1998, respectively.[22] The Caribbean islands of Anguilla, the Bahamas, and Saint Lucia now rely on tourism for more than half of all jobs, and in Malaysia, tourism is the third largest income earner, after manufacturing and oil.[23]

Yet tourism has failed to benefit many of the world's poorest nations, where political instability, poor marketing, and revenue leakage have slowed growth.[24] In some countries, as much as 55 percent of tourist income goes to local elites or is funneled back to industrial nations through foreign ownership of hotels and tour companies, according to the World Bank.[25] While Africa's share of world tourist arrivals has increased from 2 to 4 percent since 1950, its share of world receipts has actually declined, from 4 to 2 percent.[26]

The global tourism boom also poses a growing threat to the world's natural areas, from small islands to the poles.[27] Tourist transportation and infrastructure, as well as increased visitors, can bring serious pollution and habitat destruction. In 1999 alone, 21 new resort complexes were being built along Mexico's Yucatan coast, aimed at tripling tourist capacity.[28] But this intense development threatens to destroy the area's remaining mangrove forests, as well as key nesting sites for endangered sea turtles.[29]

One promising new trend is "sustainable tourism"—environmentally and socially conscious travel that can help protect natural assets as well as generate local income.[30] The rise in "green" hotels and voluntary codes of conduct for tour operators may lessen the environmental effects of the tourism boom.[31]

INTERNATIONAL TOURIST ARRIVALS AND GLOBAL TOURISM RECEIPTS, 1950–99

YEAR	ARRIVALS (million)	RECEIPTS (bill. 1998 dollars)
1950	25	13
1955	47	24
1960	69	33
1965	113	52
1970	166	66
1971	179	73
1972	189	83
1973	199	99
1974	206	99
1975	223	109
1976	229	112
1977	250	132
1978	267	152
1979	283	170
1980	286	197
1981	288	183
1982	286	162
1983	290	158
1984	317	167
1985	328	169
1986	339	201
1987	364	240
1988	395	267
1989	427	278
1990	459	322
1991	465	321
1992	503	353
1993	519	355
1994	554	378
1995	569	423
1996	600	450
1997	620	443
1998	636	439
1999 (prel)	657	449

SOURCE: World Tourism Organization, e-mails, 25 November 1999 and 22 January 2000.

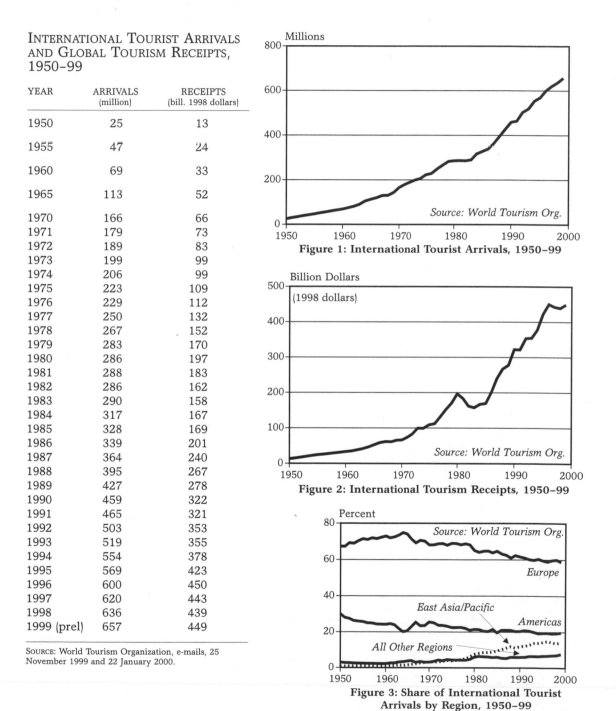

Figure 1: International Tourist Arrivals, 1950–99

Figure 2: International Tourism Receipts, 1950–99

Figure 3: Share of International Tourist Arrivals by Region, 1950–99

Transportation
Trends

Vehicle Production Increases
Michael Renner

Rising 3 percent in 1999, global passenger car production reached a record 39 million vehicles, according to estimates by Standard and Poor's DRI in London.[1] (See Figure 1.) Out of some 40 countries worldwide that manufacture automobiles, Japan, the United States, and Germany account for almost half the worldwide total.[2] (See Figure 2.) On a regional level, Western Europe is the dominant producer with 15 million cars, 38 percent of the world total; it is followed by Asia (29 percent), North America (19 percent), Eastern Europe and Russia (7 percent), and Latin America (6 percent).[3]

Worldwide, passenger car sales rose by 4.3 percent in 1999 to almost 38 million.[4] With 15 million vehicles sold, Western Europe is by far the largest market, accounting for close to 40 percent of the global total.[5] North America was second, with 9.5 million cars or 25 percent.[6] Asia continued its slow recovery from the sharp drop after the 1997 economic crisis. Its total of 6.7 million vehicles sold was equivalent to 18 percent of worldwide sales.[7] Markets in all other regions are far smaller.

The global passenger car fleet reached 520 million in 1999, according to a provisional DRI estimate.[8] (See Figure 3.) There are now 11.5 people for each car worldwide.[9] But car densities are incomparably higher in North America, Western Europe, and Japan (2–3 people per car) than, for instance, in India (224 people) or China (279 people).[10]

In the United States, so-called light trucks (sport utility vehicles or SUVs, minivans, and pickup trucks) account for a rapidly growing share of vehicle sales. From 1975 to 1999, these vehicles increased their share of new car sales from 20 percent to 46 percent.[11] Light trucks, however, are considerably less fuel-efficient than traditional passenger cars. Their growing popularity thus contributed to an erosion in fuel efficiency, from a peak of 25.9 miles per gallon (9.1 liters per 100 kilometers) in the early 1980s to 23.8 miles per gallon by 1999.[12] U.S. cars achieved 28.1 miles per gallon in 1999, but light trucks were rated a mere 20.3.[13]

Improved transmissions, fuel injection systems, and other efficiency technologies continue to be incorporated into new vehicles. But the potential efficiency gain from such technologies has been more than offset by the trend toward more powerful vehicles among cars and light trucks. Increasing vehicle weight, horsepower, and acceleration performance since 1986 have cost the equivalent of a 5-miles-per-gallon improvement.[14]

The share of transportation fuel consumed by light trucks has risen from 25 percent in 1975 to 60 percent in 1999.[15] Light trucks not only contribute disproportionately to rising fuel use, they are also far more polluting. They will be the fastest growing source of carbon emissions in the United States during this decade.[16] Concern about air pollution has led the U.S. Environmental Protection Agency to issue new rules to reduce nitrogen oxide emissions from car engines. These will be phased in between 2004 and 2009 and for the first time require that light trucks meet the same standards as regular cars.[17]

So far, the light-truck phenomenon is still largely restricted to North America. The United States and Canada produced 7.3 million passenger cars in 1999, but 15.5 million vehicles when light trucks are included.[18] In Asia, light trucks now account for one third of total vehicle production, but in Western Europe for only 11 percent.[19]

Still, sales of SUVs have also begun to pick up in Europe. Growing at a rate four times as fast as the overall car market, SUV sales there have almost doubled, from about 300,000 in 1995 to an estimated 564,000 in 1999.[20] This represents just under 4 percent of new vehicle registrations, compared with 16 percent in the United States.[21] Although European SUVs are smaller and more fuel-efficient than North American models, their growing appeal nevertheless poses a challenge at a time when combating air pollution and reducing carbon emissions are becoming ever more urgent policy goals.[22]

WORLD AUTOMOBILE PRODUCTION AND FLEET, 1950–99

YEAR	PRODUCTION (million)	FLEET
1950	8	53
1955	11	73
1960	13	98
1965	19	140
1970	23	194
1971	26	207
1972	28	220
1973	30	236
1974	26	249
1975	25	260
1976	29	269
1977	31	285
1978	31	297
1979	31	308
1980	29	320
1981	28	331
1982	27	340
1983	30	352
1984	31	365
1985	32	374
1986	33	386
1987	33	394
1988	34	413
1989	36	424
1990	36	445
1991	35	456
1992	35	470
1993	34	469
1994	36	480
1995	36	477
1996	37	486
1997	38	498
1998	38	510
1999 (prel)	39	520

SOURCES: American Automobile Manufacturers Association; Standard & Poor's DRI.

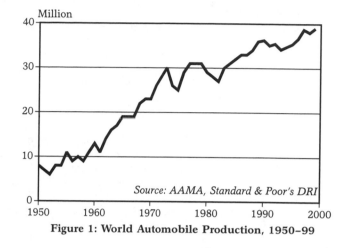

Figure 1: World Automobile Production, 1950–99

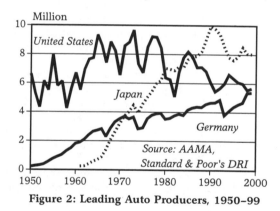

Figure 2: Leading Auto Producers, 1950–99

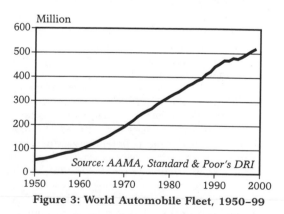

Figure 3: World Automobile Fleet, 1950–99

Global bicycle production dropped 16 percent in 1998 to 79 million units, continuing the decline in production under way since 1995.[1] (See Figure 1.) Global production now stands some 25 percent below the peak of 107 million units reached in 1995.[2] Excess inventories and sluggish demand are responsible.

Nearly all major producers reported declines, but the greatest absolute cutbacks came in Asia—traditionally the strongest producing and consuming region. China, the world's largest producer, saw output fall by 23 percent, to 23 million.[3] Chinese bicycle production has fallen by 45 percent since 1994 as trade barriers have reduced the nation's access to European and Canadian markets, and as the domestic market becomes saturated.[4]

India and the European Union, the third and fourth largest producers, saw production decline by roughly 5 percent, largely because of sluggish demand in industrial-country markets.[5] In Europe, demand for bicycles dropped by 19 percent between 1994 and 1998, due in part to a market glut produced by earlier robust sales.[6] In the United States, cycling may be on the decline. The number of people who rode a bicycle "more than once" in a calendar year fell from 53.3 million in 1996 to 43.5 million in 1998.[7]

Fifth-ranking United States reduced production by the greatest share, some 60 percent, manufacturing only 2.5 million bikes in 1998.[8] Much of U.S. production has shifted to Mexico, where lower labor costs are driving a bicycle boom.[9] Similar shifts are happening in Asia: Japanese and Taiwanese manufacturers have moved some operations to mainland China.[10]

On the bright side, sales of electric bicycles set a new record at 365,000 units for 1999.[11] (See Figure 2.) That boosted total sales since the early 1990s to well over 1 million units.[12] The trend is potentially important because electric bicycles extend biking range and allow cyclists to tackle hilly terrain that could otherwise be a deterrent to cycling.

Bicycles face a growing number of challenges in many cities. China, with some of the highest urban cycling rates in the world, has seen bike use plummet as incomes have soared and fed the demand for motorized vehicles.[13] As Chinese automobile ownership grew at roughly 15 percent a year in the 1990s, bicycles were increasingly pushed to the margins of transportation priorities.[14]

Some cities, however, are making efforts to promote cycling. Amsterdam places 250 bicycles around the city for public use in a revival of its "white bike" program of the 1960s.[15] Unlike the earlier program, which provided bicycles at no charge—resulting in heavy losses to theft within days—the new program charges 50¢ for 30 minutes of use, and features bicycles designed to deter theft and to facilitate short, crosstown trips.[16] The program is similar to one started in 1995 in Copenhagen, where more than 2,000 bicycles are now available for public use.

Such initiatives are complemented by efforts to improve cycling infrastructure in many communities. One noteworthy endeavor is the National Cycle Network, an 8,000-kilometer U.K. network scheduled to open in June 2000.[17] When completed, the network is expected to pass within 3–4 kilometers of half of the country's population.[18] Some 60 percent of trips on these paths are expected to be for commuting, shopping, and other utilitarian purposes; the remaining 40 percent are projected to be recreational.[19]

Meanwhile, Bogota is poised to give a major push to cycling. The city plans to spend more than $150 million on bicycle infrastructure and promotional efforts over nine years in an effort to boost cycling from 0.5 percent of trips to 3–4 percent.[20] More than 300 kilometers of bike routes are planned.[21]

Government interest stems from growing recognition of cycling's contributions to sustainability: bikes are efficient, inexpensive, nonpolluting, and healthy. These features prompted the Australian minister for Transport and the Minister for Health in 1999 to launch a five-year Australia Cycling campaign to combat a diverse set of societal ills: air pollution, climate change, traffic congestion, and overweight.[22]

WORLD BICYCLE PRODUCTION, 1950–98

YEAR	PRODUCTION (million)
1950	11
1955	15
1960	20
1965	21
1970	36
1971	39
1972	46
1973	52
1974	52
1975	43
1976	47
1977	49
1978	51
1979	54
1980	62
1981	65
1982	69
1983	74
1984	76
1985	79
1986	84
1987	98
1988	105
1989	95
1990	92
1991	99
1992	102
1993	104
1994	106
1995	107
1996	99
1997	93
1998 (prel)	79

SOURCES: United Nations, *The Growth of World Industry 1969 Edition*, Vol. I, *Yearbooks of Industrial Statistics 1979* and *1989 Editions,* and *Industrial Commodity Statistics Yearbook 1997*; *Interbike Directory,* various years.

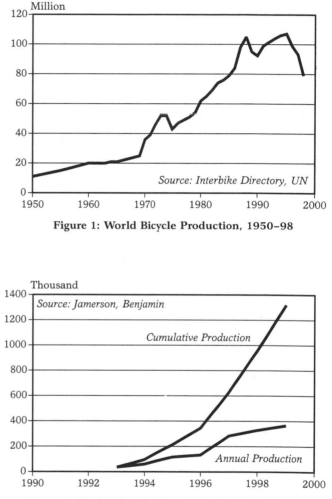

Figure 1: World Bicycle Production, 1950–98

Figure 2: World Electric Bicycle Production, 1993–99

Communication
Trends

Telephone Network Diversifies   Molly O. Sheehan

The number of fixed-line phone connections worldwide increased 7 percent to 844 million in 1998, the latest year for which data are available from the International Telecommunication Union.[1] Wireline connections increased at about this rate during the 1990s.[2] (See Figure 1.)

Conventional phone lines, however, represent an increasingly smaller share of the total network. Between 1997 and 1998, cellular phone subscriptions rose 48 percent to 319 million.[3] (See Figure 2.) Throughout the 1990s, the number of subscribers doubled every 20 months.[4] The number of new mobile subscribers edged past new fixed-line installations in 1996, and by 1998 the additional wireless connections were double the new wired connections.[5]

Although phone lines now reach every continent and calls can be placed from remote villages with a mobile phone that beams radio signals to a cellular tower or communications satellite, basic phone service is still inaccessible to many. International telephone traffic soared from about 80 billion minutes in 1997 to over 90 billion minutes in 1998, but almost three quarters of those calls originated in just 23 industrial countries.[6] Some of the new technologies and policies transforming the telecommunications industry could help to expand the network into less-served areas.

Digitization is one technological phenomenon driving change. Telephone networks have traditionally conveyed sound as analog waves. But now many types of information—not only sound, but also text, picture, or video—can be transmitted as compressed bits in the binary language of computers.[7] As a result, the lines separating traditional telephone companies from other industries are blurring.[8]

A related technological driver is growth in the capacity to transmit information, as computers have become more powerful and copper wires have been replaced by highly effective glass strands that transmit light signals. At a given instant, all of North America's long-distance telephone traffic could theoretically be carried on a single pair of these optical fibers, each the thickness of a human hair.[9] High-capacity digital connections allow communications to be provided in new ways—for instance, telephone bundled with television or Internet service. In 1999, several companies unveiled plans for wireless phones that allow users to browse the Internet.[10]

Governments are adopting new policies, both to accommodate technological change and to encourage competition. More than 150 countries introduced new telecommunications legislation or made changes to existing laws in the 1990s.[11] With the latest wave of market openings, the share of countries in which monopolies control basic phone service has dropped to 73 percent.[12]

In contrast, monopolies run mobile phone operations in only one third of countries.[13] Europe, with the most competition, also has the fastest growth in cellular phone use.[14] Current trends suggest that at some point between 2001 and 2007, the total number of mobile connections worldwide will surpass fixed-line ones.[15]

Mobile phones, most frequently used in wealthy nations, have many advantages for poorer countries. Cellular towers can be built in less time than it takes to lay cables and wires. And wireless systems may prove more durable than copper phone lines, which are often stolen for their scrap value or damaged by war.[16] The two countries that already have more cellular than fixed-line subscribers are Finland, a leader in the technology, and Cambodia, a war-ravaged nation.[17] Since 1992, when cell phones were introduced there, Cambodia has passed 31 other countries in the per capita number of phone connections.[18]

Despite their benefits, mobile phones also have drawbacks. For instance, the towers needed to transmit cellular signals disrupt the beauty of wilderness areas and urban parks. And dialing and driving can be a deadly combination; a University of Toronto study found that people who use phones while driving are four times as likely as other drivers to have an accident.[19] An unanswered question is whether the radio signals from mobile phones can harm human health.[20]

TELEPHONE LINES AND
CELLULAR PHONE SUBSCRIBERS
WORLDWIDE, 1960–98

YEAR	TELEPHONE LINES	CELLULAR PHONE SUBSCRIBERS
	(million)	
1960	89	–
1965	115	–
1970	156	–
1975	229	–
1976	244	–
1977	259	–
1978	276	–
1979	294	–
1980	311	–
1981	339	–
1982	354	–
1983	370	–
1984	388	–
1985	407	1
1986	426	1
1987	446	2
1988	469	4
1989	493	7
1990	520	11
1991	546	16
1992	574	23
1993	606	34
1994	645	55
1995	691	91
1996	738	142
1997	788	215
1998	844	319

SOURCES: ITU, *World Telecommunications Indicators
'98* (1999); ITU, *World Telecommunication Development Report 1999* (1999).

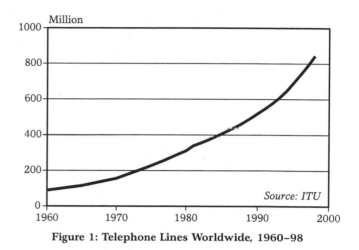

Figure 1: Telephone Lines Worldwide, 1960–98

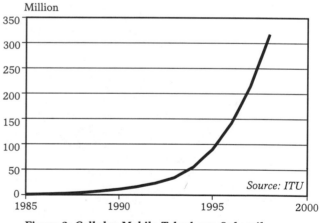

Figure 2: Cellular Mobile Telephone Subscribers
Worldwide, 1985–98

In 1999, some 72 million host computers were connected to the Internet, enabling 260 million people—over 4 percent of the planet—to go online.[1] (See Figure 1.) The Internet grew more rapidly in 1999 than during the previous four years, expanding by 67 percent.[2]

The United States is home to 111 million Internet users.[3] Although this one country claims 43 percent of the world's online population, its share of the total has shrunk from 61 percent in 1997.[4] Japan is next, with 18 million users, followed by the United Kingdom, Canada, and Germany, with about 14 million each.[5] Today, 72 million people in Europe and 47 million in Asia are online.[6]

The year 1999 marked a turning point for some developing nations, which made their debut on the list of countries with the most Internet users.[7] Brazil (with 6.9 million online), China (6.3 million users), and South Korea (5.7 million) overtook several European nations to join the top 10 list.[8]

Net access in developing countries grew 93 percent in 1999, outstripping the growth rate for the Internet as a whole.[9] Latin America's online population more than doubled in 1999, reaching 9 million.[10] Although Brazil still dominates the region, Mexico established a strong foothold when its host computer count nearly quadrupled and its total number of users reached 1 million.[11] China, which leads the tally in developing Asia, expanded its access more than fourfold, exceeding all projections.[12]

People in remote regions have eagerly taken to the Internet, capitalizing on its ability to connect them to the rest of the world. Host computer counts surged in many island nations, more than doubling in Cuba, tripling in Papua New Guinea, and growing sixfold in Madagascar.[13] Mongolia and Cambodia's host count each grew 150 percent in 1999.[14] And the mountain kingdom Bhutan has expanded its Internet base more than 15-fold each year since it went online in 1997.[15]

Several African countries went online for the first time in 1999, including Sierra Leone, Rwanda, and Malawi.[16] Yet this region's Internet infrastructure is still largely undeveloped. Its per capita host count is just 2 per 10,000

people, a tenth that of Latin America.[17] (See Figures 2 and 3.) Although ranked third in the region, Botswana's host computer count is less than that of the tiny island of Bermuda.[18] And even today, some 70 percent of the region's Internet users live in South Africa.[19]

Africa's unequal Internet distribution mirrors the global picture. While access is growing rapidly in some developing countries, most users—some 87 percent—live in the industrial world.[20] And while the Internet is ubiquitous in many affluent nations—more than 40 percent of the U.S., Canadian, and Swedish populations are online, for example— less than 1 percent of people in China, India, or Mexico have access.[21]

Although English still dominates, some 46 percent of users now surf the Internet in other languages, led by Japanese, Spanish, and Chinese.[22] Even as its audience has diversified, however, the Web has grown more concentrated: 80 percent of traffic goes to just 15,000 sites.[23] Some 2 million pages are added to the Web each day, bringing the total up to 1.5 billion pages by the end of 1999.[24]

In 1999, consumers and businesses spent $111 billion online—three times as much as in 1998.[25] The tourism industry was a big winner: 52 million Americans went online to plan their travel and make reservations.[26] And advertising on the Internet doubled in 1999, swelling to $2.8 billion.[27] Ironically, U.S. online companies spent more than $1 billion to advertise on television and in magazines during the year.[28]

The Internet can be a powerful engine of consumerism. But some analysts argue that where consumption levels are already high, the Internet might help reduce natural resource use. A Washington-based team of energy experts reports that by 2007, e-commerce could prevent the annual release of 35 million tons of greenhouse gases by reducing the need for up to 3 billion square feet of energy-consuming office buildings and malls in the United States.[29] And the Internet continues to aid environmental and social activists everywhere to communicate, share information, and campaign for sustainable development.[30]

INTERNET HOST COMPUTERS, 1981–99

YEAR	HOST COMPUTERS (number)
1981	213
1982	235
1983	562
1984	1,024
1985	2,308
1986	5,089
1987	28,174
1988	80,000
1989	159,000
1990	376,000
1991	727,000
1992	1,313,000
1993	2,217,000
1994	5,846,000
1995	14,352,000
1996	21,819,000
1997	29,670,000
1998	43,230,000
1999 (prel)	72,398,000

SOURCE: Internet Software Consortium and Network Wizards, "Internet Domain Surveys," <www.isc.org/ds/>, viewed 20 February 2000.

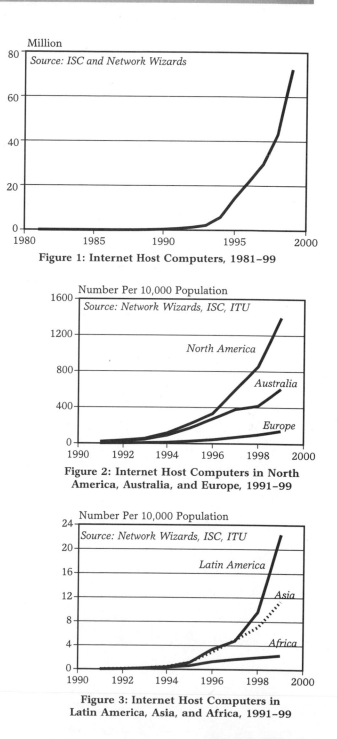

Figure 1: Internet Host Computers, 1981–99

Figure 2: Internet Host Computers in North America, Australia, and Europe, 1991–99

Figure 3: Internet Host Computers in Latin America, Asia, and Africa, 1991–99

Social
Trends

World Population Passes 6 Billion Brian Halweil

On October 12, 1999, our numbers officially reached 6 billion, double the population in 1960.[1] (See Figure 1.) Last year world population swelled by 77 million—roughly equivalent to adding another Philippines.[2] (See Figure 2.)

In addition, 1999 witnessed another population milestone when India's population surpassed 1 billion.[3] China, with 1.25 billion, still reigns as the world's most populous nation, but fast-growing India is projected to have that dubious honor by 2037.[4]

It took less time—just 12 years—to add this last billion to the planet than any previous billion, despite an annual rate of growth at its lowest level in a half-century: 1.3 percent.[5] (See Figure 3.) But because this lower rate comes on top of the largest population base ever, the world added more people in 1999 than in 1963, when the annual growth rate peaked at 2.2 percent.[6]

Even with this declining rate of growth, more young men and women—1.1 billion— are reaching reproductive age than ever before.[7] Our annual addition will still average over 70 million people each year for the next two decades before declining to roughly 30 million by 2050, when total population is expected to reach nearly 9 billion.[8]

Global population growth is concentrated in South Asia and Africa. Nearly 3 out of 10 people added to the planet in 1999 were born in the Indian subcontinent, while another 2.5 were born in Africa.[9] Most of the remainder were born in China, Latin America and the Caribbean, and Southeast Asia.

Falling death rates and rising life spans— the result of improved nutrition, sanitation, immunizations, and other public health advances—underpin the dramatic population increases in these regions. Infant and child mortality have plummeted in Latin America, Africa, and Asia since 1950, while the average life span increased from 41 to 64 years.[10] Because fertility rates—the number of children each woman bears—have not declined as fast, national populations surged in the twentieth century. (After similar but more gradual quality-of-life improvements in the nineteenth century, European and North American populations swelled, though they are now stable, or even declining in some of Europe.)

The world's poorest regions are growing most rapidly because inadequate social services and economic opportunities leave couples dependent on large families for financial security and with little power to determine their family size. Some 350 million women, nearly a third of all women of reproductive age in developing countries, have incomplete or sporadic access to safe family planning services.[11] For another 120 million women, such services simply do not exist, or cultural and religious barriers prevent their use.[12]

Fertility rates also remain high because women are not given the same opportunities as men: worldwide, women and girls make up more than two thirds of the world's illiterate population and three fifths of the poor.[13] Girls who attend school tend to delay their first child and bear fewer children overall. In Egypt, 56 percent of women with no schooling become mothers in their teens, compared with just 5 percent of women who remained in school past the primary level.[14]

Population growth can strain the capacity of governments and the environment to meet human needs. With freshwater availability essentially fixed, the number of people living in water-scarce regions will jump from 470 million to more than 3 billion by 2030.[15] In sub-Saharan Africa, where literacy and school enrollment are already well below international averages, the school-age population will expand by more than one third by then.[16]

A 1999 assessment of the progress made toward the goals laid out at the 1994 International Conference on Population and Development—universal access to family planning services, gender equity in education, and improved health care and sanitation services— found inadequate financial resources have kept these development goals beyond reach.[17] Of the estimated $17 billion needed to ensure universal access to family planning, developing nations have honored nearly 70 percent of their commitment, while industrial nations have honored only one third of theirs.[18]

WORLD POPULATION, TOTAL AND ANNUAL ADDITION, 1950–99

YEAR	TOTAL[1] (billion)	ANNUAL ADDITION (million)
1950	2.556	38
1955	2.780	53
1960	3.039	41
1965	3.345	70
1970	3.707	77
1971	3.784	77
1972	3.861	76
1973	3.937	76
1974	4.013	73
1975	4.086	72
1976	4.158	72
1977	4.231	72
1978	4.303	75
1979	4.378	76
1980	4.454	76
1981	4.530	80
1982	4.610	80
1983	4.690	79
1984	4.770	81
1985	4.851	82
1986	4.933	86
1987	5.018	86
1988	5.105	86
1989	5.190	87
1990	5.277	82
1991	5.359	82
1992	5.442	81
1993	5.523	80
1994	5.603	80
1995	5.682	79
1996	5.761	80
1997	5.840	78
1998	5.919	78
1999 (prel)	5.996	77

[1]Total at mid-year.
SOURCE: U.S. Bureau of the Census, *International Data Base*, electronic database, Suitland, MD, updated 28 December 1999.

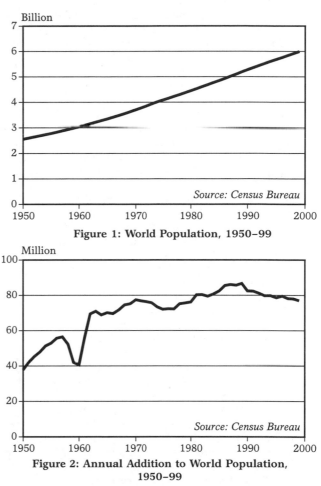

Figure 1: World Population, 1950–99

Figure 2: Annual Addition to World Population, 1950–99

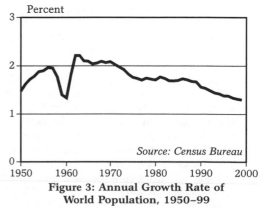

Figure 3: Annual Growth Rate of World Population, 1950–99

HIV/AIDS Pandemic Hits Africa Hardest   Brian Halweil

Thirty-six million people—nearly two thirds of them in Africa—entered the new century infected with HIV, the virus that causes AIDS.[1] Since the early 1980s, the cumulative number infected has almost reached 50 million (see Figure 1), and there are now nearly 6 million new infections each year.[2] A record number of deaths from AIDS in 1999, 2.6 million, pushed the cumulative death toll to 16 million—nearly as many people as live in New York City.[3] (See Figure 2.)

Sub-Saharan Africa remains the center of the global epidemic, with AIDS now causing one out of five deaths there each year.[4] Eastern and Southern Africa have been hit particularly hard: home to just 5 percent of the human population, these two regions contain over half of the people who are HIV-infected, and more than 60 percent of those who have died of AIDS.[5] Life expectancy in southern Africa, which climbed from 44 years in the early 1950s to 59 in the early 1990s, is expected to drop back to 45 in this decade.[6]

A newly emergent and highly virulent subtype of the virus, HIV-1C, is driving HIV prevalence to between one fifth and one third of the adult population in Botswana, Namibia, South Africa, and Zimbabwe.[7] In Africa's two most populous nations, Nigeria and Ethiopia, prevalence is growing but remains lower, at 4 and 9 percent, respectively.[8]

About 12.2 million African women are infected, compared with 10.1 million men.[9] Greater ease of male-to-female transmission, as well as greater exposure to risky sexual situations, threatens women worldwide.[10] Most of these women will unknowingly pass the virus to their babies, adding to the half-million children born infected each year in Africa.[11]

Outside of Africa, adult prevalence has topped 1 percent in only a handful of nations. The hardest hit are in Central America, the Caribbean, and Southeast Asia, and include Guyana and Cambodia, where 3 percent of the adults are infected, and Haiti, where 6 percent are HIV-positive.[12] In most of the world high infection rates are still found only in high-risk, urban populations—providing an excellent opportunity for prevention.

But continued low public awareness, the spread of intravenous drug use, and widespread unsafe sexual behavior portends explosive epidemics elsewhere.[13] For instance, although adult prevalence in Asia is a small fraction of that in Africa, the adult population is five times larger, and the number infected in Asia jumped by 25 percent in 1999.[14]

Perhaps spurred by the devastating spread of the virus in Africa, some Asian nations have pursued aggressive prevention programs, focusing on sex education, needle exchange, and reproductive health services. Thailand and the Philippines now appear to have stabilized or reduced HIV prevalence rates.[15] In the Indian state of Tamil Nadu, a mass media campaign promoting safe sex cut the rate of casual sex among factory workers in half between 1996 and 1998, while condom use rose from 17 to 50 percent.[16]

HIV/AIDS strikes hardest at the most sexually active—the young breadwinners, parents, students, and professionals who underpin household, community, and national development. In nine African nations, UNAIDS found that one fifth to one third of the children are likely to be orphaned by AIDS over the next decade.[17] By 2010, Africa could be home to 40 million AIDS orphans.[18]

In Zambia, colleges graduated 300 new teachers in 1999, but AIDS took the lives of 600 teachers.[19] In Kigali, Rwanda, 34 percent of people with a post-secondary education are infected with HIV, nearly three times the rate for the population at large.[20] In some hospitals in South Africa, AIDS patients occupy 60 percent of the beds.[21] By 2005, treatment, care, and support related to HIV/AIDS are expected to account for a third of all government health-spending in Ethiopia, more than half in Kenya, and nearly two thirds in Zimbabwe.[22]

Despite these daunting prospects, national and international responses have fallen short. In Zimbabwe, the government each month spends just $1 million on HIV/AIDS prevention and $70 million on the war in the Congo.[23] UNAIDS recently reported that the global epidemic is expanding three times faster than the international funding to prevent it.[24]

CUMULATIVE HIV INFECTIONS AND AIDS DEATHS WORLDWIDE, 1980–99

YEAR	HIV INFECTIONS (million)
1980	0.1
1981	0.3
1982	0.7
1983	1.2
1984	1.7
1985	2.4
1986	3.4
1987	4.5
1988	5.9
1989	7.8
1990	10.0
1991	12.8
1992	16.1
1993	19.7
1994	23.8
1995	28.3
1996	33.5
1997	38.9
1998	44.1
1999 (prel)	49.9

YEAR	AIDS DEATHS (million)
1980	0.0
1981	0.0
1982	0.0
1983	0.0
1984	0.1
1985	0.2
1986	0.3
1987	0.5
1988	0.8
1989	1.2
1990	1.7
1991	2.4
1992	3.3
1993	4.4
1994	5.7
1995	7.3
1996	9.2
1997	11.3
1998	13.7
1999 (prel)	16.3

SOURCE: Neff Walker, UNAIDS, Geneva, e-mail to author, 20 March 2000.

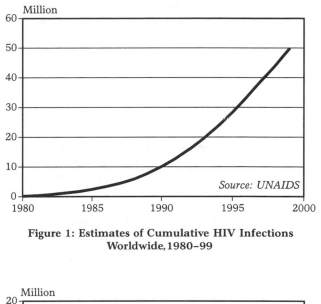

Figure 1: Estimates of Cumulative HIV Infections Worldwide, 1980–99

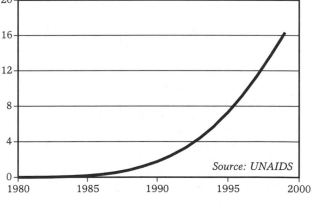

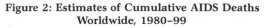

Figure 2: Estimates of Cumulative AIDS Deaths Worldwide, 1980–99

Refugee Numbers Continue Decline   Michael Renner

For the fourth consecutive year, the number of people qualifying for and receiving refugee assistance from the U.N. High Commissioner for Refugees (UNHCR) declined.[1] As of January 1999, the figure stood at 21.5 million—22 percent below the January 1995 peak of 27.4 million.[2] (See Figure 1.)

UNHCR has four categories within this total population "of concern": 11.5 million refugees (down from more than 18 million in 1992; see Figure 2), 1.3 million asylum seekers, 2.4 million recent returnees (refugees and internally displaced persons) who continue to need assistance, and 6.3 million "others of concern," including internally displaced persons.[3]

The Palestinians are the largest single refugee group, with an estimated 3.8 million.[4] The second largest group contains 2.6 million Afghanis uprooted by two decades of nearly uninterrupted internal warfare triggered by the Soviet invasion of 1979.[5] Other major sources of refugees are Iraq, Burundi, Somalia, Bosnia-Herzegovina, and Sierra Leone.[6]

Asian countries hosted the largest number of persons of concern to UNHCR—7.5 million at the beginning of 1999.[7] (See Figure 3.) African countries were second, with 6.3 million—a significant decrease from the previous year—closely followed by Europe (6.2 million).[8] Among individual countries, Iran continued to carry the heaviest burden, with almost 2 million persons in its territory.[9] It was followed by Bosnia-Herzegovina, Pakistan, Russia, the United States, and Germany, all of which hosted slightly more than 1 million people.[10]

UNHCR primarily deals with caring for international refugees—those fleeing war and persecution who have crossed a border.[11] It can look after the internally displaced only when a national government gives its consent. Yet the plight of these people is often far worse than that of recognized refugees. Not only are they not protected under international refugee law but, as the U.S. Committee for Refugees notes, "many are actively attacked by their own governments and remain largely inaccessible to outside monitors."[12]

The U.N. agency uses a rough estimate of 30 million for the number of internally displaced persons worldwide.[13] (The U.S. Committee for Refugees lists 41 countries with a combined population of internally displaced of 17–19 million as of December 1998, but notes that the total may be much higher).[14] The countries with the largest numbers of internally displaced persons are Sudan, Angola, Colombia, Afghanistan, Myanmar, and Turkey.[15] Although UNHCR's involvement with this group has increased substantially in recent years, its assistance extends only to about 5 million people, 10 percent of whom have recently returned to their homes and still receive assistance.[16]

Another group, people in "refugee-like situations," typically live in conditions similar to those of refugees although they have not received official recognition. Some are ignored or tolerated by host governments, others harassed as illegal aliens. Estimates of their numbers are fragmentary, but a tally of 30 host countries by the U.S. Committee for Refugees suggests that there were at least 5 million at the beginning of 1999.[17]

All in all, there may be some 57 million refugees and internally displaced persons, although the figure is likely considerably higher.[18] Hence, it seems that nearly one out of every 100 persons on Earth is affected.

During 1998, almost 1 million refugees returned to their home countries either with UNHCR's help or on their own.[19] By far the largest repatriations took place in Liberia, Bosnia-Herzegovina, and Afghanistan.[20] At the same time, however, tens of thousands of people elsewhere faced expulsions from their host country or were otherwise forced to return home involuntarily.[21]

There is continued concern about the erosion of the right to asylum worldwide. Instead of offering refuge abroad, western countries increasingly prefer establishing internal "safe havens." These areas are often anything but safe and there are no international rules governing them. In effect, the richer countries just shift the burden of caring for displaced people.[22]

Refugee Numbers Continue Decline

Refugees Receiving U.N. Assistance, 1961–99[1]

YEAR	TOTAL (million)
1961	1.4
1962	1.3
1963	1.3
1964	1.3
1965	1.5
1966	1.6
1967	1.8
1968	2.0
1969	2.2
1970	2.3
1971	2.5
1972	2.5
1973	2.4
1974	2.4
1975	2.4
1976	2.6
1977	2.8
1978	3.3
1979	4.6
1980	5.7
1981	8.2
1982	9.8
1983	10.4
1984	10.9
1985	10.5
1986	11.6
1987	12.4
1988	13.3
1989	14.8
1990	14.9
1991	17.2
1992	17.0
1993	19.0
1994	23.0
1995	27.4
1996	26.1
1997	22.7
1998	22.4
1999 (prel)	21.5

[1]All data are as of January of the year indicated.
SOURCE: United Nations High Commissioner for Refugees, various data series.

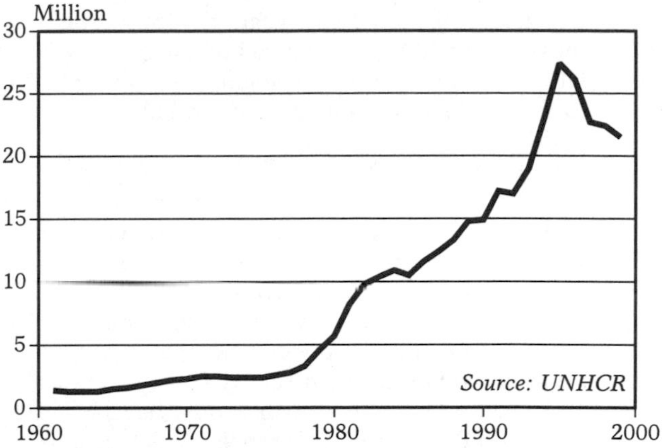

Figure 1: Refugees Receiving U.N. Assistance, 1961–99

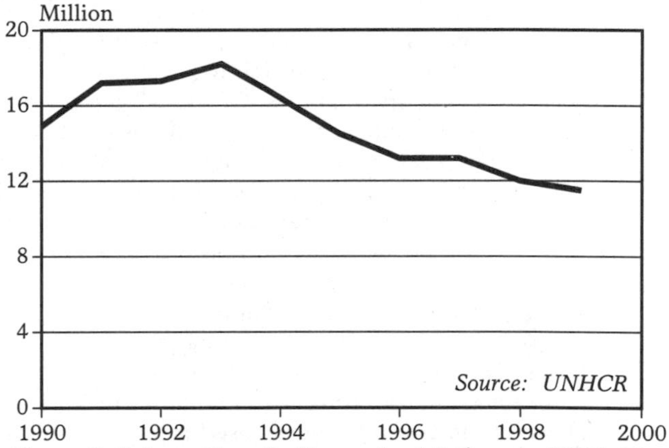

Figure 2: Internationally Recognized Refugees, 1990–99

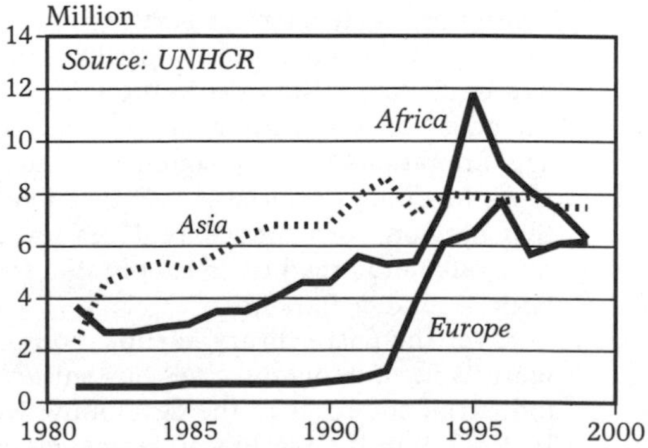

Figure 3: Refugees Receiving U.N. Assistance in Asia, Africa, and Europe, 1981–99

Urban Population Continues to Rise   Molly O. Sheehan

Between 1996 and 1999—years that mark the most recent U.N. assessments of urban population—some 200 million people were added to the world's urban areas.[1] At 2.8 billion, nearly four times as many people lived in urban areas in 1999 as in 1950.[2] (See Figure 1.) Global urban population estimates are difficult to make, as the definition of "urban" and the reliability of census data vary from country to country. The U.N. figures cited here are for "urban agglomerations," which generally include the population in a city or town as well as adjacent suburbs.

Urban population growth is outstripping rural population growth by three to one as a result of rural-to-urban migration, the natural increase of existing urban populations, and the reclassification of areas that were once rural villages. Thus this century is likely to become the first one in which most of the world's people live in cities. Of the 5.9 billion people on the planet in mid-1999, 47 percent resided in urban areas.[3] (See Figure 2.) By 2006, according to U.N. projections, half of the world will live in cities, and by 2030, three out of five people could be urbanites.[4]

Global environmental challenges such as climate change and deforestation have many urban roots.[5] Cities generate some three quarters of the carbon dioxide that is released from fossil fuel burning worldwide; a similar share of industrial timber is used in cities.[6]

Industrial nations tend to be more urbanized than developing ones and to consume a disproportionate share of Earth's resources. More than 70 percent of national populations live in cities and suburbs in the United States, Canada, Western Europe, and Japan.[7] Urban agglomerations in these countries draw heavily on far-flung resources.[8] One estimate finds that London, for instance, requires roughly 58 times its land area just to supply its residents with food and timber.[9]

Over the past century, the location of the world's most populous cities has shifted from industrial countries to the developing world. In 1900, 9 of the world's 10 largest cities were in Europe and the United States.[10] In contrast, only Tokyo, New York, and Los Angeles in the industrial world make the Top 10 list in 2000. They join Mexico City, Bombay, São Paolo, Shanghai, Lagos, Calcutta, and Buenos Aires.[11]

Population increase in urban centers of developing countries is expected to account for nearly 90 percent of the 2.7 billion people likely to be added to world population between 1995 and 2030.[12] (See Figure 3.) Some 74 percent of Latin Americans now live in cities, making the region roughly as urbanized as Europe and North America. Thus the most explosive urban growth in the future is expected in Africa and Asia, where only 30–35 percent of the population is now urban.[13]

Local environmental problems, such as water and air pollution, are worst in cities where population size or growth exceeds the capability of governments to build and maintain critical water, waste, and transportation infrastructure.[14] At least 220 million people in cities of the developing world lack clean drinking water, and 1.1 billion choke on air pollution.[15]

The battle to achieve a sustainable balance between Earth's resource base and its human energy will be largely won or lost in the world's cities. By concentrating populations, cities have a natural environmental advantage: people clustered together should in theory be able to use less materials and energy than widely dispersed populations can, and to recycle resources more easily.[16] The challenge will lie in mustering the political and financial resources needed to build the urban water, waste, transportation, and energy systems to exploit this advantage.

Urban Population Continues to Rise

WORLD URBAN POPULATION, AND SHARE THAT IS URBAN, 1950–99

YEAR	POPULATION (billion)
1950	0.750
1955	0.872
1960	1.017
1965	1.185
1970	1.357
1975	1.543
1980	1.754
1985	1.997
1990	2.280
1995	2.574
1999 (prel)	2.800

YEAR	SHARE (percent)
1950	29.7
1955	31.6
1960	33.6
1965	35.5
1970	36.7
1975	37.8
1980	39.4
1985	41.2
1990	43.2
1995	45.3
1999 (prel)	47.0

SOURCE: U.N Population Division.

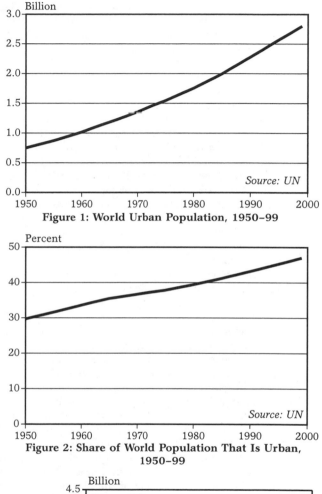

Figure 1: World Urban Population, 1950–99

Figure 2: Share of World Population That Is Urban, 1950–99

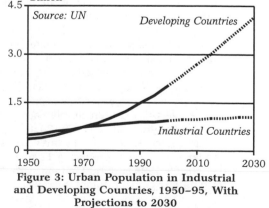

Figure 3: Urban Population in Industrial and Developing Countries, 1950–95, With Projections to 2030

Cigarette Death Toll Rising
Anne Platt McGinn

In 1999, an estimated 1.15 billion smokers worldwide lit up 14 cigarettes a day each.[1] Global cigarette production declined just two tenths of 1 percent between 1998 and 1999 to 5,485 billion, down 3 percent from an all-time high of 5,681 billion pieces in 1996.[2] (See Figure 1.) Global production per person dropped to 915 cigarettes in 1999, down 7 percent from 1996.[3] (See Figure 2.)

Despite the leveling off in total production and drop in per capita supplies, annual deaths from smoking-related causes are expected to jump from 4 million in 1998 to 10 million in 2030.[4] In 30 years, the world's leading killer will not be a disease but a consumer product.[5] And by 2020, one out of three adult deaths worldwide will result from smoking—more than the deaths expected from malaria, tuberculosis, and maternal and childhood ill-nesses combined.[6]

With 80 percent of the world's current smokers living in developing countries, where smoking rates are climbing by 3.4 percent a year, many of the illnesses and premature deaths will hit hardest in areas that can least afford to treat more than 25 known tobacco-related diseases, including heart disease, respi-ratory ailments, and cancer.[7] Worldwide, the costs of treating smoking-related illnesses are estimated at $200 billion a year, more than 10 times the tobacco industry's profits in 1999.[8]

China is the world's largest cigarette pro-ducer, accounting for 1,675 billion cigarettes or 31 percent of global supply.[9] More than 300 million men and 20 million women there currently smoke 30 percent of the planet's supply, making this country also the world's leading consumer.[10] Despite the popularity of smoking, public awareness of related health risks is extremely low: one in every two smokers in China surveyed in 1999 did not know that smoking can cause cancer.[11] In China alone, health experts predict that 2 mil-lion people will die prematurely each year from tobacco-related causes by 2020.[12]

The United States remains the world's sec-ond largest supplier, producing 12 percent of the world's 1999 supply.[13] Indonesia ranks third among global producers, contributing 4 percent of world supplies, having overtaken Japan in 1998.[14] (Japan ranks third in total cigarettes smoked, after China and the United States.)[15]

In contrast to the situation in developing countries, consumption has been declining by about 1 percent a year in many industrial nations. During the past decade, the number of smokers in Europe dropped by 10 percent, a trend that is expected to continue.[16] By 2006, the European Union will become the first region in the world to ban all cigarette advertising.[17] Per capita consumption in the United States declined to 1,634 in 1999, a 9-percent drop from 1998.[18] (See Figure 3.)

As a result of higher retail prices and taxes, tough anti-smoking laws, and increased public awareness of the hazards of smoking, U.S. tobacco companies have shifted their focus from meeting domestic demand to promoting their lethal products to consumers in Asia, Africa, and Latin America.[19] In 1998, these firms spent nearly $5 billion in advertising outside the United States.[20] When adjusted for inflation, advertising overseas has tripled in the last 20 years, according to the U.S. Federal Trade Commission.[21] Consequently, the United States continues as the world's leading cigarette exporter.[22]

Partly as a result, more than a dozen coun-tries in Latin America, Europe, and the Pacif-ic have filed lawsuits against U.S. tobacco companies seeking to recoup payment for smoking-related illnesses.[23] Modeled on U.S. cases that awarded $251 billion to state gov-ernments in 1998, these lawsuits claim simi-lar damages.[24]

And in 1999, the World Health Organiza-tion set the stage for the first legally binding international treaty on a global public health issue.[25] With more than 190 countries involved in negotiations, the Framework Con-vention on Tobacco Control will help lay the foundation for stricter regional and national measures to address the social and economic costs of smoking.[26]

WORLD CIGARETTE PRODUCTION, 1950–99

YEAR	TOTAL (billion)	PER PERSON (number)
1950	1,686	660
1955	1,921	691
1960	2,150	707
1965	2,564	766
1970	3,112	840
1971	3,165	836
1972	3,295	853
1973	3,481	884
1974	3,590	895
1975	3,742	916
1976	3,852	926
1977	4,019	950
1978	4,072	946
1979	4,214	962
1980	4,388	985
1981	4,541	1002
1982	4,550	987
1983	4,547	969
1984	4,689	983
1985	4,855	1,001
1986	4,987	1,011
1987	5,128	1,022
1988	5,250	1,026
1989	5,258	1,013
1990	5,419	1,027
1991	5,351	998
1992	5,363	985
1993	5,300	960
1994	5,478	978
1995	5,599	985
1996	5,681	986
1997	5,643	966
1998	5,609	948
1999 (prel)	5,485	915

SOURCE: USDA, *Special Report: World Cigarette Situation*, August 1999; data for 1950–58 are estimates based on U.S. data.

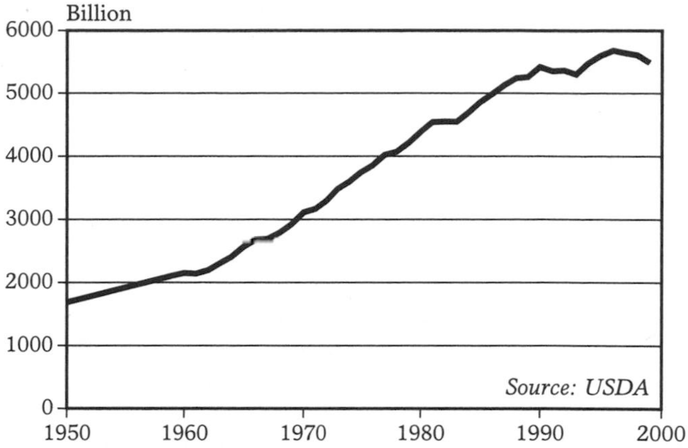

Figure 1: World Cigarette Production, 1950–99

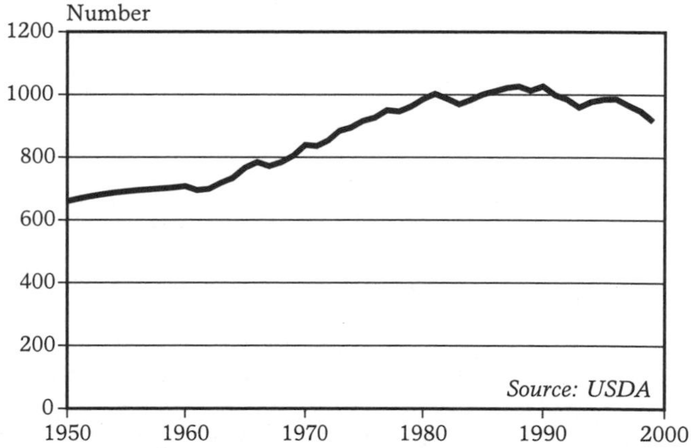

Figure 2: World Cigarette Production Per Person, 1950–99

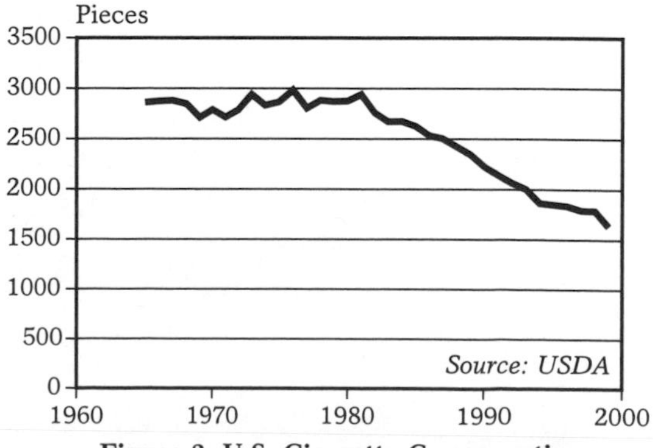

Figure 2: U.S. Cigarette Consumption Per Person, 1965–99

Military
Trends

Number of Wars on Upswing
Michael Renner

According to AKUF, a study group at the University of Hamburg, the number of wars worldwide rose to 35 in 1999—continuing an upswing since 1997.[1] (See Figure 1.) The 1999 number is still one third below the 1992 peak, when 51 wars were under way.[2] Since 1945, there have been at least 212 wars around the world.[3]

During each year from 1992 to 1997, more wars were ended than new ones started.[4] Many long-standing conflicts ended in peace agreements. This triggered tremendous hope about a future with far fewer wars. But these expectations were quickly dashed. Ethnic and religious disputes, social and economic inequities, failures of governance, environmental degradation, and other factors continue to fuel the flames of violent conflict. Conflict prevention and peacekeeping still remain marginal endeavors.[5]

A total of eight new armed conflicts broke out during 1999, according to AKUF.[6] Hostilities in Chechnya and East Timor received wide coverage in the world's media, but fighting in Aceh (Indonesia), Tripura (India), Nepal, Kyrgyzstan, the Solomon Islands, and Nigeria went largely unreported and received little attention from diplomats.[7]

For the decade from 1989 to 1998, the Conflict Data Project at the University of Uppsala, Sweden, tallies a total of 108 armed conflicts in 73 different locations.[8] The vast majority of these—92 of the 108—took place exclusively within the boundaries of a single country.[9] Another nine involved intra-state conflicts with foreign intervention.[10] Just seven wars during that decade took place between opposing states.[11]

In 1999, three such inter-state wars were active: the border war pitting Ethiopia against Eritrea, Indian-Pakistani clashes over control of Kashmir, and an on-again, off-again U.S.-British aerial bombing campaign against Iraq.[12] In addition, the Chechen and East Timorese conflicts are hybrid cases: Chechnya had de facto become an entity separate from Russia after the 1996 war, but was not internationally recognized as a sovereign state. East Timor, on the other hand, was never part of Indonesia, even though it had been occupied since 1975.

With just one exception (in Kosovo), all armed conflicts during 1999 took place in the Third World.[13] According to the Uppsala Conflict Data Project, Asia and Africa were the two regions with by far the highest number of armed conflicts—15 and 14, respectively—in 1998.[14] (See Figure 2.)

In terms of human lives lost, the costliest ongoing wars are those in Afghanistan and Sudan, with 1.9 million and 1.5 million dead, respectively.[15] They are followed by Rwanda (500,000–1 million), Angola (more than 500,000), Algeria, Burundi, Congo (formerly Zaire), Iraq, and Sri Lanka (100,000–200,000 each).[16] These conflicts are claiming primarily civilian lives—not so much directly in battle than as a result of famine and social upheaval.

The nature of war has changed tremendously. It is becoming difficult to define armed conflict, as the distinction between political and criminal violence blurs. Increasingly, fighting involves not the uniformed armed services of a state but warlords, ethnic militias, private armies, and criminal organizations. And in 87 percent of the wars active in 1998, child soldiers were used—as many as 300,000 worldwide.[17]

The violence of many contemporary armed struggles is less an expression of clear political or military objectives (such as defending a border or annexing territory) than an indication of "the social chaos borne of state failure," in the words of Ernie Regehr of Project Ploughshares.[18] An underlying factor is the failure of states to create or maintain conditions conducive to the welfare of their populations. A Project Ploughshares analysis found that 41 percent of the states in the bottom half of the U.N. Development Programme's Human Development Index in 1998 experienced war on their territories within the previous decade.[19]

ARMED CONFLICTS, 1950–99

YEAR	CONFLICTS (number)
1950	12
1955	14
1960	10
1965	27
1970	30
1971	30
1972	29
1973	29
1974	29
1975	34
1976	33
1977	35
1978	36
1979	37
1980	36
1981	37
1982	39
1983	39
1984	40
1985	40
1986	42
1987	43
1988	44
1989	42
1990	48
1991	50
1992	51
1993	45
1994	41
1995	36
1996	31
1997	29
1998	32
1999	35

SOURCE: Arbeitsgemeinschaft Kriegsursachen-forschung, Institute for Political Science, University of Hamburg.

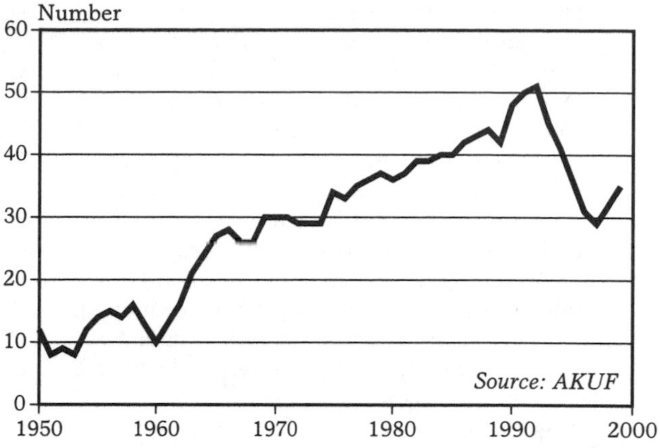

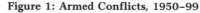

Source: AKUF

Figure 1: Armed Conflicts, 1950–99

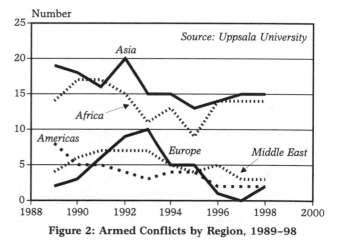

Figure 2: Armed Conflicts by Region, 1989–98

Peacekeeping Expenditures Turn Up   Michael Renner

For the first time since 1996, the United Nations expects to spend more on peacekeeping than in the previous year. For July 1999–June 2000, projected expenditures are at least $1.4 billion, and may increase to $1.8–2 billion.[1] (See Figure 1.) This is up from $860 million a year earlier, although still substantially below the peak levels of the mid-1990s. Some 14,615 soldiers, military observers, and civilian police from 84 countries served in peacekeeping missions in early December 1999 (in addition to 9,284 civilian personnel).[2] (See Figure 2.) As a few recently established missions scale up to their authorized strength, however, this figure is expected to reach about 34,000, and possibly as many as 40,000, during 2000.[3]

In recent years, U.N. peacekeeping has been overshadowed—in terms of funding and personnel—by operations run by NATO and other regional military organizations. The growing prominence of non-U.N. operations is the product of real and perceived U.N. failures in Bosnia and Somalia, an unwillingness among western governments to make adequate resources available to the United Nations, and an intent to bypass the U.N. Security Council, as in the 1999 Kosovo crisis. Operations outside U.N. purview are more concerned with a heavy-handed imposition of peace than with impartial peacekeeping, and they may serve the interests of individual countries or regional military alliances more than the interests of humanity as a whole.

During 1999, a total of 17 U.N. missions were active, including 4 new ones: a transitional administration to assist East Timor's path to independence, an interim administration for Kosovo, a peacekeeping force for Sierra Leone (upgraded from a small observer mission), and an observer mission in the Congo.[4] But 1999 also saw the end of a failed, decade-long peacemaking effort in Angola and the termination of a promising effort at conflict prevention in Macedonia.[5]

The newest missions are poised to become the largest current U.N. operations. When they reach their authorized levels, there will be 10,790 peacekeepers in East Timor, 11,100 in Sierra Leone, 4,756 in Kosovo, and 5,537 in the Congo.[6] The first three missions are working closely, though not without some friction, with non-U.N. forces. The Kosovo mission is principally a police force working in conjunction with the NATO-led KFOR military force. The Sierra Leone mission, working alongside ECOMOG (a Nigerian-led West African force), is supposed to oversee a shaky peace agreement that ended a brutal rebellion. And the East Timor mission is taking over from an Australian-led force that intervened, with U.N. blessing, to stop the rampaging of anti-independence militias.[7]

Since the beginnings of U.N. peacekeeping in 1948, a total of 53 missions have been initiated, at a cost of about $20 billion, and sent to 36 different countries, territories, or border areas.[8] Of these, 18 went to Africa, 10 to Europe, 9 to the Middle East, 8 to Asia, and 8 to Central America and the Caribbean.[9]

U.N. peacekeeping continues to struggle under the cloud of financial crisis. As of mid-December 1999, U.N. members owed the organization $1.7 billion for peacekeeping operations.[10] (See Figure 3.) The United States is still the most in arrears, with $1.05 billion in unpaid dues (61 percent of the total).[11]

Non-U.N. missions now cost seven times as much as U.N. operations and they deploy four times as many soldiers and observers. Even as Washington expresses concern about the cost of U.N. peacekeeping, the U.S. and other western governments continue to pour huge amounts of money, personnel, and equipment into operations not directed by the United Nations and far less accountable to the international community.

Two NATO-led forces, SFOR in Bosnia and KFOR in Kosovo, fielded more than 88,000 soldiers in 1999, costing an estimated $11 billion.[12] Altogether, some 35 non-U.N. missions were active that year.[13] Reliable cost and personnel data are not available for several, but collectively they deployed at least 125,000 troops and observers and their expenditures ran to at least $12 billion.[14] This is up from about 55,000 personnel and $1.4 billion as recently as 1993.[15]

U.N. PEACEKEEPING EXPENDITURES, 1986–99

YEAR	EXPENDITURE (mill. 1998 dollars)
1986	338.3
1987	325.9
1988	348.2
1989	797.0
1990	558.3
1991	567.4
1992	1,991.4
1993	3,359.1
1994	3,584.3
1995	3,527.1
1996*	1,338.0
1997*	1,002.0
1998*	860.0
1999*	1,435.7

*July to June of following year.
SOURCES: U.N. Department of Peacekeeping Operations; Office of the Spokesman for the U.N. Secretary-General.

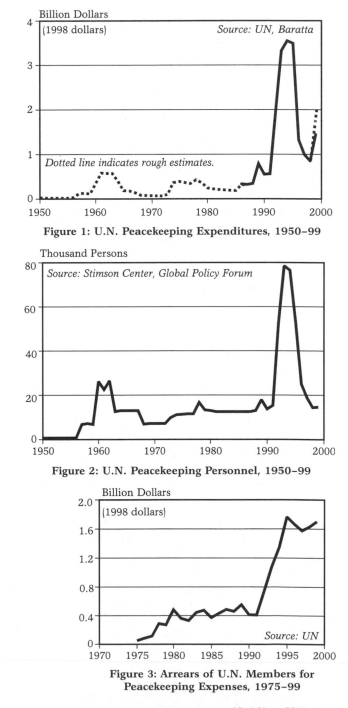

Figure 1: U.N. Peacekeeping Expenditures, 1950–99

Figure 2: U.N. Peacekeeping Personnel, 1950–99

Figure 3: Arrears of U.N. Members for Peacekeeping Expenses, 1975–99

Part TWO

Special Features

Environmental Features

Transgenic Crop Area Surges

Brian Halweil

The global area planted to transgenic crops jumped by 44 percent between 1998 and 1999, from 27.8 million hectares to 39.9 million.[1] The area has grown 23-fold since 1996, the first year of large-scale commercialization, when just 1.7 million hectares were planted.[2] But 99 percent of the current global transgenic area is found in just three nations—the United States, Argentina, and Canada; 72 percent of the global area is in the United States alone.[3] (See Table 1.)

More than half of the American soybean and cotton crops and one third of the corn crop is modified; 90 percent of the Argentine soy crop is modified, as is nearly two thirds of the Canadian rapeseed (canola) crop.[4] Because these three nations are dominant food exporters, much of the corn, soy, canola, and cottonseed on the world market is transgenic.[5] Elsewhere, public concerns or ongoing government evaluation of these crops has delayed widespread planting.

Also known as genetically engineered or genetically modified crops, transgenics often contain genes from viruses, bacteria, animals, and other organisms with which the crop could not reproduce naturally. Traditional breeding, in contrast, involves humans crossing only closely related plant species.

Dozens of crops—from apples to lettuce to wheat—have been modified and are near commercialization, though only transgenic varieties of soybean, corn, cotton, canola, potato, squash, and papaya are currently grown commercially.[6] Of these seven crops, soybeans and corn account for 54 percent and 28 percent of the global transgenic area, respectively, while cotton and canola share most of the remainder, with nearly 9 percent each.[7]

The transgenic crops currently being grown have been engineered to resist spraying of herbicides, to churn out the insecticide produced by the soil bacterium *Bacillus thuringiensis* (Bt), or to do both. In 1999, herbicide-resistant varieties of soy, corn, cotton, and canola were planted on 71 percent of the global transgenic area, while Bt-corn and Bt-cotton were sown on 22 percent.[8] Corn and cotton varieties that both produce Bt and resist herbicides were planted on the remaining 7 percent—a sevenfold increase in the use of such "trait-stacked" varieties.[9] These traits offer large-scale, industrial farmers reduced production costs or increased ease of crop management by lowering the need to scout for pests, cutting labor costs, allowing a shift to cheaper chemicals, and generally simplifying pest control—which explains the exceptionally rapid adoption of transgenics in a few nations.[10]

Public resistance to transgenic crops spread from Europe to the rest of the world in 1999, galvanized by several risk assessment studies and high-profile lawsuits. A hotly debated study suggested that genetically engineered potatoes

TABLE 1: GLOBAL AREA OF TRANSGENIC CROPS, 1999

COUNTRY	1999 (million hectares)	SHARE OF GLOBAL ACREAGE (percent)
United States	28.7	72
Argentina	6.7	17
Canada	4.0	10
China	0.3	1
Australia	0.1	< 1
South Africa	0.1	< 1
Mexico	< 0.1	< 1
Spain	< 0.1	< 1
France	< 0.1	< 1
Portugal	< 0.1	< 1
Romania	< 0.1	< 1
Ukraine	< 0.1	< 1
Total	39.9	100

SOURCE: Clive James, *Global Review of Commercialized Transgenic Crops: 1999 (Preview)* (Ithaca, NY: International Service for the Acquisition of Agri-Biotech Applications, 1999).

could damage the immune system and internal organs of rats, bolstering concerns that transgenic foods might induce allergies or toxic reactions in humans.[11] Scientists also showed that the pollen produced by Bt-corn killed Monarch butterfly larvae in the lab, while another study reported that the toxin produced by Bt-corn could accumulate—in its active form—in the soil for extended periods of time.[12] Both studies raised concerns about possible unanticipated or untested ecological impacts.

Monsanto and AstraZeneca bowed to public pressure by deciding in 1999 not to commercialize so-called terminator technologies that would have rendered seeds sterile and prevented the age-old practice of seed saving—though research on these technologies seems to be proceeding nonetheless.[13] And a lawsuit brought in December charged that Monsanto Company, the leading producer of transgenic seeds, had not adequately tested its seeds before commercialization and that it was trying to monopolize the seed supply through gene patenting.[14]

Such events rippled throughout the food chain and the investment community. Most major food manufacturers and retailers decided to remove transgenic ingredients from their products sold in Europe.[15] Several Japanese and American food companies, including Asahi, Heinz, Gerber, and Frito-Lay, followed suit.[16] Japan, South Korea, Australia, Mexico, the members of the European Union, and other nations began to draft laws requiring mandatory labeling of food products containing transgenic ingredients.[17]

The shift in public perception cost U.S. agriculture hundreds of millions of dollars, as exports to Europe plummeted.[18] Top commodity handlers, including Archer Daniels Midland and A.E. Staley, began to discount transgenic crops because of the greater financial risk.[19] And several major players in the biotech industry, including Novartis, Astra-Zeneca, and the newly merged Pharmacia Upjohn and Monsanto, spun off their ailing agricultural units.[20]

A series of studies in 1999 indicated that—in contrast to claims by biotech proponents—the adoption of such crops is not reducing the use of harmful pesticides, and in some cases has increased it by making spraying easier.[21] (Bt-cotton stands as the exception: studies have indicated as much as a 12-percent reduction in insecticide applications on Bt-cotton in the United States compared with conventional cotton systems.)[22] Other observers have also noted that Bt- and herbicide-resistant crops keep farmers firmly rooted on the pesticide treadmill and vulnerable to pesticide resistance.[23]

In early 2000, scientists in Switzerland announced that they had developed a variety of transgenic rice that was enhanced with beta-carotene—dubbed "golden rice" for its yellow color and intended to alleviate debilitating vitamin A deficiency throughout Asia.[24] This breakthrough was heralded as evidence that transgenic crops could help reduce malnutrition. Issues of cultural acceptance remain, however, as well as concern that such a technology does not sufficiently address the poverty and overly monotonous diet at the root of the deficiency.[25]

The first international treaty regulating trade in transgenic products was established in January 2000, allowing nations to bar imports of transgenic crops and other organisms based on environmental, human health, and social risks, even in the face of scientific uncertainty over such risks.[26] This biosafety protocol also requires that shipments of agricultural commodities indicate whether they "may contain" transgenic ingredients. The protocol was more ambiguous on its relationship with the World Trade Organization, setting the stage for future trade disputes.[27]

Global planting in coming years will largely be affected by the evolution of public sentiment in the United States—the largest producer and consumer of transgenics. Labeling bills have been introduced in the U.S. Congress, and U.S. regulatory agencies are reviewing their oversight of transgenics.[28] Faced with uncertain domestic and export markets, surveys show that U.S. farmers plan to scale back the area planted to transgenics by 15–25 percent in 2000.[29]

Organic Farming Thrives Worldwide   Brian Halweil

Driven by rising consumer demand and growing dissatisfaction with conventional farming practices, the organic agriculture industry is soaring. A recent U.N. survey found that farmers in at least 130 countries on all continents produce organic food commercially.[1] The International Federation of Organic Agriculture Movements (IFOAM)—the primary international standard setting and lobbying body for the organic industry—has more than 750 members in 107 nations today, up from 5 members in 3 nations in 1972, with most of the new members in the developing world.[2] Total global organic area is now estimated at more than 7 million hectares, while the market for organic food has swelled to an estimated $22 billion a year.[3]

The term "organic" describes a system of farming that prohibits the use of synthetic pesticides and artificial fertilizers, and instead relies on ecological interactions to raise yields, reduce pest pressures, and build soil fertility. Diverse planting patterns, frequent rotations, and attraction of beneficial insects, for instance, would all be "organic" means of pest control.[4]

The European Union (EU) leads the global organic explosion, with a 35-fold expansion in organic area since 1985—an average annual growth rate of 30 percent.[5] (See Figure 1.) At nearly 4 million hectares in 1999, organic area accounts for roughly 3 percent of total EU agricultural area, while the retail market for organic products has hit some $7.3 billion.[6] In several European nations, including Sweden, Finland, Switzerland, and Italy, 5–10 percent of total agricultural area is now organic.[7] In Austria, 13 percent of the farmland is organic, with the share reaching half in some provinces.[8]

But Australia is the nation with the most organic area, with 1.7 million certified organic hectares, raising mostly organic range-fed beef for export to Japan—where the organic market is now worth $3.5 billion.[9] In the United States and Canada, the organic area grew 15–20 percent each year during the 1990s, and now stands at roughly 550,000 and 1 million hectares—0.2 and 1.3 percent of the respective nation's cropland.[10] Retail sales of organic produce and products in North America have also grown 20 percent annually since 1989, and were estimated at $10 billion in 1999.[11]

These swelling markets for organic products have sometimes been driven by policies to promote organic farming, and at other times by market forces. For example, 80 percent of the growth in EU area occurred in the last six years, spurred by the 1993 establishment of a common EU definition for "organic"—which is integral to boosting consumer awareness—and subsequent policies to support conversion to organic farming, such as subsidies in the early years of conversion and organic farming research at agricultural universities.[12]

In contrast, growth in the United States has come despite little government support. A study by the Organic Farming Research Foundation found that less than one tenth of 1 percent of U.S. Department of Agriculture research projects in 1995 had any relevance for organic agriculture.[13] And an aborted 1997 effort by the government to set federal organic standards would actually have weakened the industry by permitting transgenic seeds, confined livestock operations, and other inputs and practices never before considered organic.[14]

A series of food safety, ecological, and other troubles associated with the conventional food sector, including the "mad cow" scare in the United Kingdom, has also inspired a fierce market demand for organic.[15] Among the British, recent concerns over genetically engineered crops (which are not permitted in organic production) caused an avalanche of consumer inquiries about organic and a parallel flood of farmer applications for conversion.[16] Since 1996, organic area in the United Kingdom surged 10-fold, from 50,000 to 500,000 hectares.[17]

Statistics for the developing world are spotty, although anecdotal evidence points to rapid growth, especially for export markets.[18]

In Argentina, the total area devoted to organic production jumped 7,000 percent since 1992 to an estimated 350,000 hectares today.[19] Argentina is expected to export more than $100 million of organic products in 2000.[20] An Export Promotion of Organic Products from Africa project was started in 1995 in Mozambique, Tanzania, Uganda, and Zimbabwe, with the dual goal of addressing rural poverty and resource conservation.[21] At least 7,000 small farmers in Uganda—up from 220 in 1995—now produce about 10 percent of the organic cotton on the world market.[22]

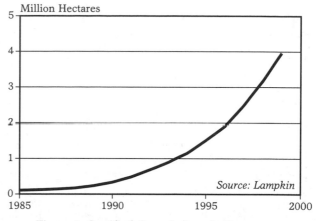

Figure 1: Certified Organic Area in European Union, 1985–99

Local organic markets are also emerging in the developing world. In Egypt, where tea drinking is a daily ritual, the top national brand is Sekem's certified organic tea.[23] And in Cuba, a nationwide shift to organic farming includes an estimated 30,000 urban gardens, which are a principal source of fresh produce for city dwellers.[24]

Several recent studies have indicated that yields from organic production are comparable to conventional systems, especially over the long term.[25] Combined with the price premiums that organic produce often fetches in the market, organic systems are thus generally more profitable for the farmer. A recent study of organic grain and soybean production in the U.S. Midwest found that organic systems were often more profitable even without the price premium because of the lower input costs, a greater diversity of products being sold, and greater yield stability in bad-weather years—all pluses for subsistence farmers in ecologically sensitive areas.[26]

Organic farming has also demonstrated a wide range of ecological benefits, including reduced soil erosion, improved soil health, and reduced groundwater contamination.[27] A joint declaration from IFOAM and the World Conservation Union–IUCN supported organic agriculture based on its role in the conservation of biodiversity and habitat.[28] Direct human health benefits from organic farming include the reduced risk of pesticide poisoning for farm workers, as well as lower exposure to pesticide residues for consumers of organic foods—the primary driver of the booming organic baby foods market.[29]

With more and more nations drafting organic standards, setting organic area goals, and supporting organic agriculture, the prospects for further growth are bright.[30] In January 1999, the U.N. Food and Agriculture Organization said it will begin providing information on organic farming and trade, give related institutional and policy support, and explore the feasibility of organic farming for improving food security and natural resource use in the developing world.[31]

On the current trajectory, as much as 30 percent of total EU farmed area could be organic by 2010.[32] In the United States, revised national organic standards released in March 2000 will spur the domestic market.[33] And most major food manufacturers and retailers in Europe and North America have introduced their own organic product lines, while several large apparel companies, including The Gap, Levi's, and Patagonia, have begun to purchase organic cotton.[34]

Groundwater Depletion Widespread   Sandra Postel

As farmers have increasingly turned to underground sources of water to irrigate their crops, the overpumping of groundwater is causing water tables to decline beneath vast areas of agricultural land. Based on the best data available, farmers are collectively overpumping regional groundwater sources by at least 160 billion cubic meters a year—the amount of water used to produce nearly one tenth of the world's current grain supply.[1] The problem, moreover, is worsening and represents one of the largest threats to future food production.

Groundwater, stored in underground geologic formations called aquifers, is in many ways an ideal source of water. Whereas large river-based canal systems often deliver water unreliably and lose a significant portion of water stored in reservoirs to evaporation, groundwater usually can be tapped whenever a farmer needs it. And because water is stored underground, none is lost to evaporation. Affordable and decentralized, groundwater wells have proliferated mainly through private farmer investments—in stark contrast to the large government subsidies often doled out for large dams and river diversion projects.

In India, the number of groundwater wells climbed from 4 million in 1951 to 17 million in 1997, allowing the area irrigated by groundwater to climb sixfold—to 36 million hectares.[2] This rapid growth was a major contributor to food production gains during the Green Revolution, which combined high-yielding seeds, fertilizer, and water to boost land productivity. Groundwater now accounts for half of India's total irrigation water use, as well as half of the water used by cities and industries.[3]

Groundwater development has fueled food production gains in the three other top irrigators as well. In China, the number of irrigation wells increased more than 20-fold between 1961 and the mid-1980s.[4] In Pakistan, where 80 percent of all cropland is now irrigated, the number of wells rose from some 25,000 in 1964 to nearly 360,000 in 1993.[5]

In the United States, third in total irrigated area (after India and China), a groundwater boom occurred during the last half of the twentieth century. Farmers in California stepped up their pumping beneath the rich soils of the Central Valley, turning this region into the nation's premier fruit and vegetable basket. And in the Great Plains, farmers began to tap on a large scale one of the planet's greatest aquifers—the Ogallala. Spanning portions of eight states, the Ogallala covers some 453,000 square kilometers, and, prior to exploitation, held 3,700 cubic kilometers of water—a volume equal to the annual flow of more than 200 Colorado Rivers.[6] (A cubic kilometer is a billion cubic meters.) Today, the Ogallala alone waters one fifth of U.S. irrigated land.[7]

Like any renewable resource, groundwater can be tapped indefinitely as long as the rate of extraction does not exceed the rate of replenishment. But just like a bank account, a groundwater reserve will dwindle if withdrawals exceed deposits. Few governments have established and enforced rules and regulations to ensure that groundwater sources are exploited at a sustainable rate. As a result, a classic "tragedy of the commons" has unfolded: acting in their own self-interest, individual irrigators pump as much water as they desire, which collectively depletes the resource.

In India, the situation became so severe over the last decade that the Supreme Court directed one of the nation's premier research centers to study it. The National Environmental Engineering Research Institute found that "overexploitation of ground water resources is widespread across the country," and that water tables in critical agricultural areas were dropping "at an alarming rate."[8] Nine Indian states are now running major water deficits, which in the aggregate total just over 100 billion cubic meters (bcm) a year.[9] (See Table 1.)

Northern China is also running a chronic water deficit, with groundwater overpumping of some 30 bcm a year.[10] Water tables have been dropping 1–1.5 meters a year under much of the north China plain, which produces some 40 percent of China's grain.[11] The projected 2025 water deficit for the Hai and

TABLE 1: WATER DEFICITS IN KEY COUNTRIES AND REGIONS, MID-1990s

COUNTRY/ REGION	ESTIMATED ANNUAL WATER DEFICIT (billion cubic meters per year)
India	104.0
China	30.0
United States	13.6
North Africa	10.0
Saudi Arabia	6.0
Minimum Global Total	163.6

SOURCE: Various references cited in the text and author's estimates.

Yellow river basins roughly equals the volume of water needed to grow 55 million tons of grain—14 percent of the nation's current annual grain consumption.[12]

In the United States, several decades of heavy pumping have depleted the Ogallala aquifer by 325 bcm, a volume equal to the annual flow of 18 Colorado Rivers.[13] More than two thirds of this depletion has occurred in the Texas High Plains.[14] Annual net depletion of the Ogallala averages about 12 bcm a year.[15] Particularly in its southern reaches, the Ogallala gets very little replenishment from rainfall, so pumping the large volumes of water needed to grow cotton and corn inevitably diminishes the supply.

Irrigation in the arid regions of North Africa and the Arabian Peninsula depends heavily on fossil aquifers—groundwater reserves that formed thousands of years ago, when local climates were wetter than at present. Saudi Arabia, for example, is estimated to have some 2,000 cubic kilometers of 10,000- to 30,000-year-old water stored in aquifers down to a depth of 300 meters.[16] Since fossil reserves get negligible replenishment from rainfall, they are essentially nonrenewable: pumping water from them depletes the supply just as pumping from an oil reserve does.

How farmers and governments respond to declining groundwater supplies will greatly influence future global crop production. At some point, pumping costs climb too high or well yields drop too low to continue business-as-usual. Farmers can then choose to take land out of production, eliminate a harvest or two, switch to less water-intensive crops, or adopt more-efficient irrigation practices. Improving efficiency is the only option that can sustain crop production while lowering water use. Yet virtually everywhere groundwater depletion is occurring, efforts to raise efficiency—including, for example, the use of drip irrigation systems, precision sprinklers, laser-leveling of fields, and better irrigation scheduling—pale in comparison to the scale of the problem.[17]

More than food security is at stake. Because groundwater often sustains rivers, wetlands, and lakes, the overpumping of aquifers can cause serious ecological harm. In the Upper Guadiana catchment in Spain, for instance, a 30–40 meter decline in the water table has dried out valuable wetlands.[18] Rivers that depend on groundwater for their base flow can run dry when water tables drop too far, decimating fisheries. And overpumping of coastal aquifers can reverse the hydraulic gradient between land and sea, causing saltwater to invade freshwater sources. Israel, Florida in the United States, and the Indian state of Gujarat are among the areas battling the contamination of drinking water supplies by seawater.[19] Finally, if aquifers compress when water is removed from their pores, the overlying land can subside and cause considerable damage to buildings and infrastructure. Bangkok, Mexico City, and Venice are among the major cities faced with this problem.

No government has yet adequately tackled the issue of groundwater depletion, but it is at least getting more attention. The first big hurdle is overcoming the out-of-sight, out-of-mind syndrome and the human tendency to deny problems that seem too big or difficult to confront.

Groundwater Quality Deteriorating Payal Sampat

Even as our dependence on groundwater has grown over the past 50 years, the quality of this vital resource has been deteriorating in several parts of the world. The pollution of the world's aquifers, which are vast underground stores of water, represents a serious threat to global freshwater availability.

Aquifers store most of the world's unfrozen fresh water—some 97 percent—and provide drinking water to almost a third of the planet's people.[1] Over a billion residents of Asia alone depend on groundwater for drinking.[2] And groundwater has been central to the global expansion in irrigated agriculture. For instance, aquifers water more than half of irrigated land in India and 43 percent in the United States.[3] Groundwater also replenishes streams, lakes, wetlands, and other surface water bodies; it provides the base flow for some of the world's great rivers, including the Yangtze and the Mississippi.[4]

But the capacity of groundwater to sustain people and ecosystems is under enormous threat.[5] (See Table 1.) Across the United States and in parts of Asia, Latin America, and Europe, human activities are sending massive quantities of chemicals into aquifers, causing irreversible damage to freshwater supplies. Pesticides and fertilizers that run off from farms and front lawns, petrochemicals that drip out of leaky storage tanks, chlorinated solvents and heavy metals discarded by industries, and radioactive wastes from nuclear operations are among the principal contaminants of groundwater.

Nitrates are commonly found in shallow aquifers near farms and urban areas. In Sri Lanka, 79 percent of groundwater samples contained nitrates at levels above the World Health Organization (WHO) drinking water guideline of 10 milligrams per liter.[6] A study in northern China found nitrates in groundwater at five times this guideline in more than half of the 69 locations tested.[7] Much of this contamination was the result of excessive fertilizer applications: the Chinese scientists found that crops in the region used only 40 percent of the nitrogen that was applied.[8]

Consumed at high concentrations, nitrates can cause suffocation in infants and have been implicated in digestive tract cancers.[9]

Petrochemicals, many of which are known or suspected human carcinogens, are among the most pervasive groundwater contaminants in oil-dependent countries.[10] The U.S. Environment Protection Agency found in 1998 that 100,000 underground storage tanks for petroleum were leaking beneath gas stations and factories across the country.[11]

Some 30–80 million people in Bangladesh and the Indian state of West Bengal are drinking water containing arsenic at levels between 5 and 100 times the WHO guideline.[12] Scientists believe that aquifer sediments in the Ganges delta are naturally rich in arsenic, but that residents were not exposed to the heavy metal until the 1970s, when their water supply was switched from surface to groundwater.[13]

Fluoride is another naturally occurring contaminant.[14] Fluoride is an essential nutrient, but consuming it at high concentrations can cause dental problems and crippling neck and back damage.[15] WHO estimates that 70 million people in northern China and 30 million in northwestern India are drinking water with high fluoride levels.[16]

In some cases, aquifers are polluted by effluents intentionally sent there. For instance, some 60 percent of U.S. liquid hazardous waste—34 billion liters of solvents, heavy metals, and radioactive materials each year—is injected into deep underground wells for disposal.[17] Although the effluents are sent below the deepest sources of drinking water, some wastes have managed to enter drinking water supplies in parts of Florida, Ohio, Oklahoma, and Texas.[18]

Once persistent pollutants get into groundwater, the damage is virtually irreversible.[19] In part this is because the water remains in aquifers for very long periods: on average, the residence time for groundwater is 1,400 years, in comparison with just 16 days for river water.[20] So pollutants accumulate, unlike in rivers and streams, where they are more easily flushed out. For this reason, chemicals used several decades ago are still found in groundwater. The pesticide DDT, for instance, still lingers in U.S. groundwater even though its use was banned in the late 1960s.[21]

It may take several years before we discover the aftereffects of today's chemical-dependent, throwaway economy, in part because of the unique nature of aquifers. Few countries track the health of these reservoirs—their enormous size and remoteness make them extremely expensive to monitor. And because groundwater moves very slowly—less than a foot a day in some cases—damage done to

aquifers may not be detected for decades.[22]

The pollution of groundwater strains the availability of an already limited resource. On every continent, many major aquifers are being drained faster than their natural rate of recharge, resulting in an annual overdraft of at least 160 billion cubic meters.[23] In some cases, the overpumping causes the aquifer's sediments to compact, permanently shrinking its storage capacity. In California's Central Valley, this loss is equal to more than 40 percent of the combined storage capacity of all human-made reservoirs across the state.[24]

TABLE 1: SELECTED CHEMICAL THREATS TO GROUNDWATER

THREAT	SOURCES	HEALTH AND ECOSYSTEM EFFECTS	PRINCIPAL REGIONS AFFECTED
Pesticides	Runoff from farms, backyards, golf courses; landfills	Organochlorines linked to reproductive and endocrine damage in wildlife; organophosphates and carbamates linked to liver and nervous system damage and cancers	United States, Eastern Europe, China, India
Nitrates	Fertilizer runoff; manure from livestock operations; septic systems	Restricts amount of oxygen reaching brain, which can cause death in infants ("blue-baby syndrome")	Mid-Atlantic United States, north China plain, Western Europe, Northern India
Petro-chemicals	Underground petroleum storage tanks	Benzene and other petrochemicals can be cancer-causing even at a low exposure	United States, United Kingdom, parts of former Soviet Union
Chlorinated Solvents	Metals and plastics degreasing; fabric cleaning, electronics and aircraft manufacture	Linked to reproductive disorders and some cancers	Western United States, industrial zones in East Asia
Arsenic	Naturally occurring	Nervous system and liver damage; skin cancers	Bangladesh, Eastern India, Nepal, Taiwan
Fluoride	Naturally occurring	Dental problems; crippling spinal and bone damage	Northern China, northwestern India

SOURCE: See endnote 5.

Ice Cover Melting Worldwide   Lisa Mastny

From the polar regions to high mountain glaciers, Earth's ice cover is melting at an astonishing rate.[1] (See Table 1.) Global ice melt accelerated rapidly during the 1990s—the warmest decade on record.[2] Scientists suspect that the enhanced melting is related to the unprecedented release of greenhouse gases by humans during the past century.[3]

The ice-covered polar regions are warming faster than the planet as a whole, and melting rapidly.[4] The Arctic sea ice, covering an area roughly the size of the United States, has lost an average of 34,300 square kilometers—an area larger than the Netherlands—each year since 1978.[5] But the ice has thinned even faster than it has shrunk. Between 1958–76 and the mid-1990s, the average thickness dropped from 3.1 meters to 1.8 meters, a decline of some 40 percent.[6]

The massive Antarctic ice cover, which

Table 1: Selected Examples of Ice Melt Around the World

NAME	LOCATION	MEASURED LOSS
Arctic Sea Ice	Arctic	Has shrunk by 6 percent since 1978, with a 14-percent loss of thicker, year-round ice. Has lost 40 percent of its thickness in less than 30 years.
Greenland Ice Sheet	Greenland	Has thinned by more than a meter a year on its southern and eastern edges since 1993.
Columbia Glacier	United States	Has retreated nearly 13 kilometers since 1982. In 1999, retreat rate increased from 25 meters per day to 35 meters per day.
Wilkins Ice Shelf	Antarctica	Lost nearly 1,100 square kilometers in area in early March 1999. Ice front is back 35 kilometers from previous extent.
Tasman Glacier	New Zealand	Has thinned by more than 100 meters in the past century. (Overall, New Zealand glaciers shrank some 26 percent between 1890 and 1998.)
Gangotri Glacier	India	Average rate of retreat is now 30 meters a year, compared with 18 meters a year between 1935 and 1990 and 7 meters a year between 1842 and 1935.
Caucasus Mountains	Russia	Glacier volume has declined 50 percent in the past century.
Tien Shan Mountains	China	22 percent of glacial ice volume has disappeared in the past 40 years.
Mt. Kenya	Kenya	Largest glacier has lost 92 percent of its total mass since the late 1800s. Some 40 percent of this decline has occurred since the 1960s.
Alps	Western Europe	Overall glacial extent has declined 30–40 percent since 1850. Ice has lost 50 percent of its mass in the past century.
Glacier National Park	United States	Since 1850, the number of glaciers has dropped from 150 to fewer than 50. The remaining glaciers could disappear completely in 30 years.
Upsala glacier	Argentina	Has retreated 60 meters per year over the last 60 years, and rate is accelerating.

Source: See endnote 1.

averages 2.3 kilometers in thickness and represents 91 percent of Earth's ice, is also melting—although there is disagreement over how quickly.[7] One study estimates that the Western Antarctic Ice Sheet (WAIS), the smaller of the continent's two ice sheets, has retreated at an average rate of 122 meters a year for the past 7,500 years—and is in no near danger of collapse [8] But other studies suggest that the sheet may break more abruptly if melting accelerates. They point to signs of past collapse, as well as to fast-moving ice streams within the sheet that could speed ice melt, as evidence of potential instability.[9]

For now, most Antarctic melting has occurred on the continent's edges, on the ice shelves that form when the land-based ice sheets flow into the ocean and float.[10] Within the past decade, three ice shelves have crumbled: the Wordie, the Larsen A, and the Prince Gustav.[11] Two more, the Larsen B and the Wilkins, are in full retreat and expected to break up soon, having lost more than a seventh of their combined area since late 1998—a loss the size of Rhode Island.[12] Icebergs as big as Delaware have also broken off Antarctica, posing threats to open-water shipping.[13]

Outside the poles, most ice melt has occurred in mountain and subpolar glaciers, which respond much more rapidly to temperature changes.[14] As a whole, the world's glaciers are now shrinking faster than they are growing, and losses in 1997–98 were "extreme," according to the World Glacier Monitoring Service.[15] Scientists predict that up to a quarter of global mountain glacier mass could disappear by 2050, and up to half by 2100—leaving large patches only in Alaska, Patagonia, and the Himalayas.[16] Within the next 35 years, the Himalayan glacial area alone is expected to shrink by one fifth.[17]

The disappearance of Earth's ice cover would significantly alter the global climate—though the net effect remains unknown. Ice reflects large amounts of solar energy back into space and helps cool the planet.[18] When ice melts, however, this exposes land and water surfaces that retain heat—leading to even more melt and creating a feedback loop that accelerates the overall warming.[19] But excessive ice melt in the Arctic could also cause cooling in parts of Europe and the eastern United States, as the influx of fresh water into the North Atlantic may disrupt the northward flow of the warming Gulf Stream.[20]

As mountain glaciers shrink, large regions that rely on glacial runoff for water supply could experience severe shortages.[21] The Quelccaya Glacier, the traditional water source for Lima, Peru, is now retreating by some 30 meters a year—up from only 3 meters a year before 1990—posing a threat to the city's 10 million residents.[22] And as the Himalayas melt, the glacier-fed Indus and Ganges rivers are expected to initially swell and then fall to dangerously low levels, affecting the crops and drinking water of the estimated 500 million people who live along their tributaries in northern India.[23]

Rapid glacial melting can cause serious flood damage in heavily populated regions such as the Himalayas.[24] In Nepal, a glacial lake burst in 1985, sending a wall of water rushing 90 kilometers down the mountains, drowning people and destroying houses.[25]

Large-scale ice melt would also raise sea levels and flood coastal areas, currently home to half the world's people.[26] Over the past century, melting in ice caps and glaciers has contributed on average about a fifth of the estimated 10–25 centimeter (4–10 inch) global sea level rise.[27] But ice melt's share in sea level rise is increasing, and will accelerate if the larger ice sheets crumble.[28] Antarctica alone is home to 70 percent of the planet's fresh water, and collapse of the WAIS, an ice mass the size of Mexico, would raise sea levels by an estimated 6 meters—while melting of both Antarctic sheets would raise them nearly 70 meters.[29] (Loss of the Arctic sea ice or of the floating Antarctic ice shelves would have no effect on sea level because these already displace water.)[30]

Wildlife is already suffering as a result of global ice melt—particularly at the poles, where polar bears, penguins, seals, and other creatures depend on food found at the ice edge.[31]

Stresses on Amphibians Grow Ashley T. Mattoon

Global amphibian decline emerged as a serious scientific possibility in 1989 at the first World Congress of Herpetology in Canterbury, England.[1] Biologists at the meeting began comparing notes and discovered that amphibian populations all over the world seemed to be disappearing. In many cases, amphibians were vanishing in remote protected areas, where there was no direct evidence of human influence.[2]

Amphibians—frogs, toads, salamanders, and the lesser-known "legless salamanders" called caecilians—are the world's oldest terrestrial vertebrate class, but because most of them are inconspicuous, relatively little is known about them. At the time of the Canterbury conference, for example, it was not clear whether scientists were observing natural population fluctuations or a more insidious global phenomenon.[3] Today, a wealth of new evidence has convinced nearly all specialists that a catastrophic decline is indeed occurring.[4] Large-scale disappearances have been documented in places as diverse as Costa Rica, Australia, and the United States. (See Table 1.)[5]

Virtually every major type of environmental stress has been identified as a cause for the decline of one amphibian species or another. Perhaps the most obvious reason has been the loss or degradation of habitat. In the United Kingdom, populations of all six native amphibian species have dropped precipitously due to the loss of breeding ponds—in some places, 80 percent of these ponds have been filled in the last 50 years.[6] Habitat degradation is thought to be the primary reason the Arroyo toad of southern California is missing from 75 percent of its historic range.[7] And in the national forests of western North Carolina, it is estimated that clearcutting results in the demise of nearly 14 million salamanders a year.[8]

Another leading culprit is epidemic disease.[9] The chytrid fungus, for example, has recently been linked to catastrophic die-offs in Australia, Costa Rica, Panama, and the United States.[10] Iridoviruses have been found responsible for the deaths of amphibians in the United Kingdom and the United States.[11]

In 1999, new evidence from Costa Rica suggested that the disappearance of the famous Golden Toad could be the first documented extinction due to modern climate change.[12] The toad, last seen in 1989, inhabited a cloud forest atop a mountain range. Scientists found that a long-term rise in sea surface temperatures caused the mountains' cloud bank to lift, so it was no longer depositing the amount of moisture that the Golden Toads depended on.[13] The resulting drier conditions are thought to be a primary reason behind the toad's disappearance.

Other identified causes of amphibian decline include the intentional or accidental introduction of non-native predators or competitors, ultraviolet radiation, acid rain, and agricultural pollution.[14] Rarely is only one of these many stresses acting in isolation, however. It is more likely that many disappearances have been due to a combination of threats.

For example, a pathogenic fungus may not be lethal under normal conditions, but if immune systems are weakened due to changes in climate or increased exposure to ultraviolet radiation, amphibians would be more vulnerable.[15] In some cases, the combined effect of non-native species introductions and epidemic disease have been lethal.[16] Some scientists hypothesize that the international trade in aquarium fish is to be blamed for the arrival of the chytrid fungus in Australia.[17] In the Ural Mountain region of Russia, the combined influence of species introduction and industrial pollution has caused the demise of many natives: *Rana ridibunda*, an introduced frog, has been able to displace native species because it is more tolerant of pollution.[18]

Amphibian decline is probably bad news for many other organisms. Many scientists argue that amphibians are important bioindicators—a sort of barometer of Earth's health, since they are more sensitive to environmental stress than other organisms.[19] For instance, amphibians rely on both aquatic and terrestrial

TABLE 1: LARGE-SCALE LOSSES OF AMPHIBIANS

LOCATION	SPECIES	STATUS	SUSPECTED CAUSE
Montane areas of eastern Australia	14 species of frogs, including the southern day frog and the gastric brooding frog.	Sharp population declines since the late 1970s. Four species are thought to be extinct.	Parasitic fungus, possibly introduced through international trade in aquarium fish and amphibians.
Monteverde region of Costa Rica	20 species of frogs and toads (40 percent of total frog and toad fauna), including the Golden Toad.	Disappeared after synchronous population crashes in 1987. Missing throughout 1990–94 surveys.	Climate change combined with other factors, such as parasites.
Yosemite region of California	5 of the region's 7 frog and toad species—including the mountain yellow-legged frog and the foothill, yellow-legged frog.	Severe declines—one species has disappeared entirely, another has declined to a few small populations.	Overall cause unknown. Introduced predatory fish combined with drought-induced loss of habitat contributed to the decline of some species.
Montane areas of Puerto Rico	12 of 18 endemic amphibian species	Three may be extinct, the others are in decline or at risk.	Unknown. Possibly climate change.

SOURCE: See endnote 5.

environments and are therefore vulnerable to stresses in both realms. They are vegetarians as juveniles and carnivores as adults, which can make them especially susceptible to changes in the food web. They have thin, permeable skin that can readily absorb contaminants from water, air, or soil. They do not have fur or feathers, and their eggs are not enclosed by protective shells that would shield them from ultraviolet radiation or pollution.

And amphibian decline itself is likely to become a form of ecological degradation, since amphibians play a critical role in many eco-systems. In some habitats, the biomass of amphibians can exceed that of all other vertebrates combined.[20] Amphibians are often vital links in food webs—they eat plants and animals and they are also a major food source for birds, reptiles, fish, and mammals. Many of the creatures that amphibians eat are often thought of as pests to humans—mosquitoes, for example.

Amphibians are an incredibly diverse group of organisms; in fact, there are more species of amphibians than there are of mammals.[21] The rich and largely unknown diversi-ty undoubtedly embodies a great deal of useful information, so the loss of amphibians is a tragedy for society as well. Many important medicines have been discovered from chemicals found in amphibians, including painkillers and treatments for victims of burns and heart attacks.[22] For generations, indigenous tribes in Ecuador have used a secretion from the skin of a local frog that produces a painkiller 200 times more powerful than morphine without negative side effects.[23] A U.S. pharmaceutical company is currently developing a drug modeled on the active chemical found in the secretion.[24]

As the list of documented losses grows every day, important conservation efforts are gaining steam. An essential objective is to improve understanding of the status of amphibians through the collection of long-term data. A task force of the World Conservation Union–IUCN has been compiling data from monitoring programs around the world and plans to release a comprehensive summary of the declining amphibian phenomenon in 2002.[25]

Endocrine Disrupters Raise Concern
Anne Platt McGinn

Recent research has confirmed that a growing number of synthetic chemicals—in everything from pesticides to industrial compounds—are hormonally active compounds.[1] Some chemicals can mimic, disrupt, or otherwise interfere with the body's network of hormones and receptors, known as the endocrine system.[2] (This regulates many biological processes in the body, including reproduction, metabolism, and development from conception to death.)[3] Other endocrine disrupters are associated with delayed intellectual development and immunological effects.[4]

Global production of synthetic chemicals has skyrocketed since the 1930s, from near zero to nearly 300 million tons in the late 1980s.[5] It continues to grow today, although estimates of global totals are unavailable.[6] At least 75,000 different chemicals are now used in pesticides, pharmaceuticals, plastics, and countless industrial and consumer products.[7]

Beginning in the late 1980s, mounting evidence showed that a subset of synthetic chemicals can cause long-term reproductive problems at small levels of exposure. In the early 1990s, toxicologists identified 45 pesticides and industrial chemicals as known or suspected endocrine disrupters.[8] By 1998, other chemicals were added, bringing the total to nearly 60, including several heavy metals and industrial compounds.[9] (See Table 1.) One recent estimate identifies 250 such compounds.[10] The list is likely to grow as testing and screening continue on thousands of compounds found in everything from pharmaceuticals to consumer products.[11]

Some endocrine disrupters, such as DDT and PCBs, persist in the environment and the food chain.[12] Even though these chemicals have been banned for several decades, they continue to collect in sediments and food. They amass in body fat and are carried up the food chain from prey to predator, bioaccumulating at ever higher concentrations. When people and animals eat fish, meat, milk, and other animal products, they may consume high levels of these chemicals, which can then stay in their bodies for long periods.

Sometimes exposure to endocrine-disrupting chemicals can be more direct, such as when phthalates leach from a teething ring directly into an infant's mouth.[13]

Much of the evidence of the health effects of endocrine disrupters comes from animals in the wild. After a series of marine mammal die-offs in the Baltic, Mediterranean, and North Seas during the 1970s and 1980s, researchers looked at whether chemical pollution might be responsible for the reproductive failures. In one experiment, scientists found that females who ate clean fish bred normally with males 83 percent of the time, whereas only one third of the females who consumed highly contaminated fish mated successfully.[14] Findings of hormonal and immunological harm have been confirmed in other species, from eagles in the Great Lakes to alligators in Florida and fish in the United Kingdom.[15]

One of the first signs of the impacts of endocrine disrupters on humans came during the 1950s and 1960s. Nearly 1 million pregnant women in the United States took an artificial hormone, diethylstilbestrol (DES), to prevent spontaneous abortions.[16] The drug had severe side effects. DES daughters suffered from fertility problems, abnormal pregnancies, reproductive organ malfunctions, immune system disorders, and higher rates of a rare vaginal cancer typically only seen in women over 50.[17] DES sons reported cryptorchidism (undescended testicles), abnormal semen, and hypospadias (abnormal urethral openings).[18]

More subtle are the developmental and neurological effects that are linked to some endocrine disrupters. In the United States, more than 200 children whose mothers ate PCB-contaminated salmon and lake trout from Lake Michigan while pregnant suffered from impaired intellectual development.[19] By sixth grade, these children lagged up to two years behind their classmates in reading ability and word comprehension.[20] Researchers confirm similar effects among children exposed to PCBs and dioxins in the Netherlands.[21]

Some scientists maintain that endocrine disrupters are linked to significant drops in sperm counts over the past 60 years in industrial countries. In 1992, a report in the *British Medical Journal* noted a 50-percent drop in sperm production between 1938 and 1991 among European and American men.[22] Other studies have cited a rash of related male reproductive health problems in industrial countries since the 1960s, including higher incidences of testicular cancer, cryptorchidism, and hypospadias.[23]

But the debate over male reproductive health is highly controversial. Sperm counts collected in Brazil, Hong Kong, India, Israel, Kuwait, Nigeria, and Thailand show no clear trends since monitoring began in 1978, and even an increase in some areas.[24] Moreover, some urban populations in Europe and the United States experienced no drop in sperm counts.[25] Several long-term projects are now under way to analyze the trends and determine what role—if any—endocrine disrupters may be playing.

Although the process of simply identifying these chemicals and understanding their effects is gaining momentum, more than 1,000 new chemicals are introduced to the global market each year without any prior testing for endocrine effects.[26]

TABLE 1: SELECTED ENDOCRINE DISRUPTERS BY CHEMICAL CATEGORY, PRODUCTION DATA, AND HEALTH EFFECTS

CHEMICAL	GLOBAL OR NATIONAL PRODUCTION	HEALTH EFFECTS
Pesticides		
Atrazine	377 tons in United States in 1996	Spermatoxicity, birth defects, low birth weight, spontaneous abortions
DDT	3 million tons since 1942	Weakly estrogenic, feminization
Endosulfan	817 tons in United States in 1995	Male infertility, hormonal effects
Industrial Chemicals		
Dioxins (byproduct from waste incineration, paper and pulp making, industrial processes)	10.5 tons International Toxic Equivalency of dioxins and furans combined, 1995	Interferes with enzymes and hormones; affects reproduction and sex organs; carcinogenic
PCBs (used in electrical transformers, hydraulic fluids)	1–2 million tons since 1929	Mimic estrogens, interfere with thyroid hormones; decreased birth weight and delayed brain development
Phthalates (used in plastics for industrial. medical, and household uses)	Most abundant synthetic chemicals; 454,000 tons a year in United States alone	Hormonal effects, male infertility, birth defects, spontaneous abortions
Heavy Metals		
Lead (paint, construction materials, electronics, ceramics)	2.7 million tons in 1995	Male and female infertility, spontaneous abortions; neurological effects, developmental delays, birth defects
Manganese (gasoline, coal-fired power plants)	22 million tons in 1995	Low birth rates, slowed fetal development, neurological toxicity

SOURCE: Ted Schettler et al., *Generations at Risk: Reproductive Health and the Environment* (Cambridge, MA: The MIT Press, 1999); DDT production data from Paul Johnston, David Santillo, and Ruth Stringer, "Marine Environmental Protection, Sustainability, and the Precautionary Principle," *Natural Resources Forum*, May 1999; lead and managanese production data from Gary Gardner and Payal Sampat, *Mind Over Matter: Recasting the Role of Materials in Our Lives*, Worldwatch Paper 144 (Washington, DC: Worldwatch Institute, December 1998).

Paper Recycling Remains Strong   Janet N. Abramovitz

The volume of paper recovered worldwide more than tripled between 1975 and 1997—going from 35 million to nearly 110 million tons.[1] At the same time, the share of paper used that is recycled or recovered—the wastepaper recovery rate—moved from approximately 38 percent to more than 43 percent.[2] By 2010, global use of recovered paper is expected to reach 177 million tons, with a projected recovery rate of 45 percent.[3]

Paper recovery rates vary dramatically among countries. (See Table 1.) Legislation to aggressively reduce solid waste in Germany has resulted in recovery rates of nearly 72 percent.[4] In Japan, the world's second largest paper producer, limited domestic resources and a shortage of waste disposal options have encouraged the heavy use of recovered paper.[5]

Both mandatory laws and voluntary targets have been very successful in expanding recovery and recycling. In the 1970s and early 1980s, only about one quarter of wastepaper was recovered in the United States.[6] Due to a variety of laws and private initiatives (such as banning paper in landfills, establishing curbside recycling programs, and issuing mandates for recycled content paper), recovery rates there rose to 46 percent by 1997.[7]

A 1994 European Union Directive targeted a recovery rate of 50–65 percent for packaging waste by 2001, and a new directive calls for a nearly two-thirds reduction in the amount of biodegradable material (such as paper) sent to landfills.[8] Combined with expanded recycling programs, these laws will reduce waste and increase paper recovery. The Netherlands is on the way to meeting its goal of recovering more than 72 percent of the paper sold inside its borders by 2001.[9]

In many countries, the primary motivation for increasing recovery rates has been reducing the flow of waste. The volume of waste generated in many industrial countries has grown substantially in recent decades, more than doubling in the United States alone since 1960.[10] Paper accounts for the largest share of municipal solid waste in many industrial countries. In the United States, for example,

TABLE 1: PAPER RECOVERY AND USE IN TOP 10 PAPER-PRODUCING COUNTRIES, 1997

COUNTRY	TOTAL RECOVERED	EXPORTS	IMPORTS	RECOVERY RATE[1]	UTILIZATION RATE[2]
		(1,000 metric tons)		(percent)	(percent)
United States	40,909	6,823	630	46	40
Japan	16,546	312	362	53	54
China	8,760	4	1,618	27	38
Canada	3,110	688	2,088	47	24
Germany	11,279	2,739	918	72	59
Finland	607	49	84	35	5
Sweden	1,323	193	559	55	17
France	4,270	750	998	41	49
South Korea	4,530	0	1,452	66	72
Italy	2,784	53	926	31	49
World	128,725	16,460		43	44

[1]Total recovered paper volume divided by apparent paper and paperboard consumption. [2]Total recovered paper consumption divided by paper and paperboard production.
SOURCE: Miller Freeman, Inc., *International Fact and Price Book 1999* (San Francisco: 1999).

paper makes up 39 percent (by weight) of this waste.[11] Even though almost half of that is now diverted for recycling, some 44 million tons are discarded each year—more than all the paper consumed in China.[12]

Because used paper is traded between nations, recovery rates do not necessarily indicate the amount of old paper a country actually uses to produce new paper. In fact, 15 percent of all recovered paper entered world trade in 1997.[13] Although Sweden recovers over half of what it consumes, it is such a large producer and exporter of paper that the relative contribution of recovered paper to overall paper production is only 17 percent.[14] In the United States, the largest exporter of used paper, rates for utilization of wastepaper remained close to 23 percent between 1965 and 1985.[15] But by 1997, they reached 40 percent, a level not seen since the 1940s.[16]

Although recycling has slowed growth in the demand for wood pulp, it has served more as a supplement than as a substitute for virgin fiber. Global paper consumption has been increasing so rapidly that it has over-whelmed gains made by recycling. So while the amount of material recovered has increased sevenfold since 1961 and its share of the fiber supply has nearly doubled (from 20 percent in 1961 to 38 percent in 1997), the total volume of virgin wood pulp and paper consumed and waste generated continues to rise, overtaking these important successes.[17]

Expanding the collection and reuse of old paper is one of the most promising ways of reducing the pressure to cut more trees, easing overburdened waste disposal systems, and cutting energy use and pollution. Producing new paper from old is efficient: for each ton of used paper, nearly a ton of new can be produced—far more efficient than the 2–3.5 tons of trees used to make 1 ton of virgin paper.[18] And because recycled paper has already been processed, far less energy and chemicals are required during reprocessing, just 10–40 percent of the energy consumed for virgin pulping, for example.[19] Recycling can make use of the "urban forest"—the huge supply of wood

and paper waste generated in cities.

Some grades of waste paper, such as old corrugated boxes and newspapers, are more widely recycled than others, and there are well-developed markets for pulping them to make new like products. In the United States, for example, over 70 percent of old corrugated containers have been recycled since 1995.[20]

Other grades, such as office paper, have lower recovery rates, and very little of what is collected is used to make new office paper. Instead, it is downgraded for other uses such as cardboard because of the variety of inks used and the demand for ultra-bright white office paper. In fact, more than 90 percent of the printing and writing paper made in the United States is from virgin fiber, and only 6–7 percent from recycled.[21] In the United Kingdom, which has scarce raw materials and imports two thirds of its newsprint, only 40 percent of its old newspapers are recycled.[22] The Newspaper Publishers Association there recently issued a report confirming that expanded recycling would provide considerable environmental benefits and improve the industry's competitiveness.[23]

In recent years there have been dramatic advances in the quality of recycled papers, thanks to innovations in processing (such as enzymatic deinking to remove stubborn toner inks).[24] The most common standards for judging writing papers—opacity and brightness—are easily met by today's recycled papers. The strength of recycled paper is also on a par with virgin paper—a concern for printers because breaks in large paper rolls can be very costly. And many consumers have stated a strong preference for recycled.

The potential for using old paper to provide a steady stream of fiber for new paper has yet to be fully exploited. Today's 43-percent recovery rate is far below the 70 percent or more of old paper that could be recycled.[25] Expanding paper recovery efforts to reach more businesses and homes could help achieve these levels. So, too, could eliminating widespread subsidies to the virgin fiber industry, landfills, and incinerators that put recycling at an economic disadvantage.[26]

Environmental Treaties Gain Ground Hilary French

Five new environmental agreements were forged in 1999, bringing the list of international environmental accords to nearly 240. (See Figure 1.) More than two thirds of these pacts have been reached since the first U.N. conference on the environment was held in Stockholm in 1972.[1]

As environmental diplomacy matures, negotiators are increasingly strengthening existing treaties rather than devising entirely new accords. In keeping with this broader trend, all the agreements made in 1999 built on existing treaties.

Many environmental treaties are regional, involving issues such as management of shared river systems and air corridors or the protection of migratory bird species. This was the case for three of the agreements finalized in 1999, two of which were negotiated under the auspices of the U.N. Economic Commission for Europe (ECE).

The ECE cooperated with the World Health Organization's Regional Office for Europe to broker a protocol on water and health that was adopted in London in June. This addition to a 1992 ECE convention on transboundary waterways aims to reduce water-related disease by requiring signatories to provide adequate sanitation and safe drinking water.[2]

And in November, negotiators reached agreement in Gothenburg, Sweden, on a groundbreaking protocol to the 1979 ECE Convention on Long-Range Transboundary Air Pollution. The new accord takes an innovative multifaceted approach, setting emissions reduction targets for four different pollutants. Specific targets vary by country, but the overall goal is to reduce Europe's sulfur emissions 63 percent from 1990 levels by 2010, nitrogen oxides emissions by 41 percent, volatile organic compounds by 40 percent, and ammonia emissions by 17 percent.[3]

In the Caribbean region, 16 countries agreed in October to a protocol on reducing land-based sources of marine pollution; this agreement falls under the 1983 Convention for the Protection and Development of the Marine Environment of the Wider Caribbean Region.[4]

In addition to many developments at the regional level, the last few decades have seen steady progress toward developing international rules governing the "global commons," including the atmosphere, the ocean, and biological diversity.[5] In line with this trend, two of the accords reached in 1999 are global in scope.

In early December, negotiators agreed to a Beijing Amendment to the 1987 Montreal Protocol on ozone depletion that adds a new chemical, chlorobromomethane, to the list of controlled substances and strengthens limits on another, hydrochlorofluorocarbons.[6] And in mid-December, governments adopted a protocol to the Basel Convention on the hazardous waste trade that put in place a system of liability and compensation for accidents during waste shipment.[7]

Judging from the number of treaties, environmental diplomacy appears to have been a spectacular success. And many of these accords have in fact yielded important results. Among other achievements, air pollution in Europe has declined dramatically as a result of

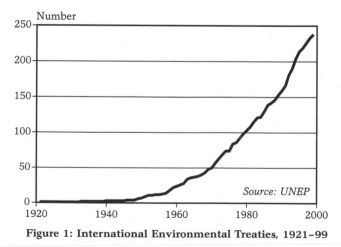

Figure 1: International Environmental Treaties, 1921–99

the 1979 treaty on transboundary air pollution; global chlorofluorocarbon (CFC) emissions have dropped by nearly 90 percent as a result of the 1987 Montreal Protocol on ozone depletion; and mining exploration and development have been precluded in Antarctica for 50 years under a 1991 accord.[8]

Yet even as the number of treaties climbs, the condition of the biosphere continues to deteriorate. Carbon dioxide levels in the atmosphere have reached record highs, scientists are warning that we are in the midst of a period of mass extinction of species, the world's major fisheries are depleted, and water shortages loom worldwide.

The main reason that many environmental treaties have not yet turned around the environmental trends they were designed to address is because the governments that created them permitted only vague commitments and lax enforcement.

One implementation tool that many environmental treaties do rely on heavily is transparency, including detailed reporting of actions taken at the national level to put agreements into practice. If this information is made freely available, then other countries as well as nongovernmental organizations can use it to shame countries into compliance.[9]

But governments often fail to provide secretariats with accurate, complete, and timely information. For example, only 63 percent of the parties of the Convention on Biological Diversity had submitted the required reports as of December 1999.[10]

The mini-institutions set up by each treaty play a key role in the implementation process. At a minimum, each treaty spawns a conference of the parties (COP) and a secretariat. The COPs are regular meetings of treaty members; they provide an opportunity to strengthen the agreement and review problems in implementation. Secretariats are the small offices set up to service these meetings of governments. Environmental conventions also commonly include scientific bodies, which provide advice on new scientific and technological information relevant to the implementation of the accord.[11]

Despite their importance, governments all too often give secretariats limited resources and authority. For instance, the secretariats generally do not have the wherewithal or authority to verify the information that governments are supposed to supply on implementation efforts. A typical secretariat has fewer than 20 staff and an annual budget of $2–11 million—a drop in the bucket compared with the budgets of the agencies charged with implementing domestic environmental laws in major countries.[12]

Although transparency can be a powerful enforcement tool, tougher medicine is sometimes required. One option is to use trade restrictions to encourage countries to participate in international environmental accords, or to abide by those they have signed on to. The Montreal Protocol on ozone depletion, for example, restricts signatories from trading in CFCs and products containing them with countries that have not joined in the accord. These provisions are widely credited with helping to bring about near universal participation in this landmark treaty.[13] But the use of trade levers as an enforcement tool is controversial, in part because of possible conflicts with the rules of the World Trade Organization.

The punitive approach embodied by penalties and sanctions has its place, but it is not always appropriate or effective. Shortages of money and governmental capacity can make it difficult for countries to comply with treaty requirements, particularly in the developing world. A critical issue for the success of most treaties is whether adequate funding and technical assistance is made available to help developing countries implement them.[14]

The last few decades of the twentieth century were a period of unprecedented activity in environmental diplomacy. The challenge for the early years of the twenty-first century will be to build upon this legacy, primarily by strengthening existing accords and ensuring that they are put into widespread practice around the world.

Economic Features

Environmental Tax Shifts Multiplying

David Malin Roodman

As of early 2000, nine countries had raised taxes on environmental harm and used the revenue to pay for cuts in taxes on income. West European governments have enacted all of these modest, revenue-neutral "tax shifts" in order to penalize destructive activities such as carbon emissions and pesticide use while encouraging constructive activities such as work and investment. Denmark and the Netherlands made two tax shifts each, and the United Kingdom will almost certainly make another soon, so that the total should soon reach a dozen tax shifts.[1] (See Table 1.)

The idea that governments should tax pollution goes back more than 80 years, to the work of Cambridge don Arthur C. Pigou.[2] He argued that if consumers and businesses had to pay for the environmental damage they did—the sulfur-corroded buildings and the lost days of work from smog-induced lung disease—they would seek ways to cut pollution in order to save money. If a pollution tax made high-sulfur coal pricier than low-sulfur, then steel companies might burn cleaner coal simply to save money. Or consumers today might unscrew their Edison-vintage incandescent bulbs in favor of efficient compact fluorescents.

Thus environmental taxes should allow governments to push an economy toward environmental soundness without taking on the impossible task of planning all the major changes that will be needed in how people

TABLE 1: TAX SHIFTS FROM WORK AND INVESTMENT TO ENVIRONMENTAL DAMAGE

COUNTRY, FIRST YEAR IN EFFECT	TAXES CUT ON	TAXES RAISED ON	REVENUE SHIFTED[1] (percent)
Sweden, 1991	Personal income	Carbon and sulfur emissions	1.9
Denmark, 1994	Personal income	Motor fuel, coal, electricity, and water sales; waste incineration and landfilling; motor vehicle ownership	2.5
Spain, 1995	Wages	Motor fuel sales	0.2
Denmark,1996	Wages, agricultural property	Carbon emissions from industry; pesticide, chlorinated solvent, and battery sales	0.5
Netherlands, 1996	Personal income and wages	Natural gas and electricity sales	0.8
United Kingdom, 1996	Wages	Landfilling	0.1
Finland, 1996	Personal income and wages	Energy sales, landfilling	0.5
Germany, 1999	Wages	Energy sales	2.1
Italy, 1999	Wages	Fossil fuel sales	0.2
Netherlands, 1999	Personal income	Energy sales, landfilling, household water sales	0.9
France, 2000	Wages	Solid waste; air and water pollution	0.1
United Kingdom, 2001[2]	Wages	Energy sales to industry	0.3

[1]Expressed relative to tax revenue raised by all levels of government.   [2]Proposed by the government but not enacted as of March 2000.
SOURCE: See endnote 1.

use resources—where they live, how they move about, and how they make everything from bottles to buildings.

Starting in the 1970s and 1980s, European analysts pointed to another benefit: governments could use the revenues from environmental taxes to cut conventional taxes on income, wages, profits, sales, trade, and built property—all essentially taxes on work and investment.[3] Since the total tax burden would remain stable under this scheme, the overall economic cost would probably be quite small or nil—and that is before counting the economic benefits of a healthier environment. Most prominent among tax shifting advocates was German environmentalist Ernst Ulrich von Weizsäcker, whose 1992 book, *Ecological Tax Reform*, became a best seller among environmentalists.[4]

Sweden was first to take up tax shifting. For many decades, "the Swedish Model" had blended socialism and capitalism to produce steady economic growth and a strong social safety net. But by the late 1980s, a recession, high taxes, unemployment, and mounting environmental problems led to critical rethinking and major reforms, including tax reform.[5] Starting in 1991, Sweden simplified and cut the personal income tax.[6] To cover the revenue gap, it raised $2.4 billion a year from new taxes on carbon and sulfur dioxide emissions.[7] Between 1989 and 1995, as power plants and factories cut energy use, switched to lower-sulfur grades of oil, and installed more smokestack "scrubbers," Sweden saw a 40-percent drop in sulfur emissions—a third of which the government attributes to the new tax.[8]

Then in the mid-1990s came the first multi-country wave of tax shifts, with the largest ones again occurring in Nordic nations. In 1994, Denmark began phasing in a 2.5-percent revenue shift from taxes on personal income to charges for energy and water use, as well as for waste incineration, landfilling, and car and truck ownership.[9] In 1996, it initiated a second shift, while the Netherlands and Finland also made changes.[10]

Increasingly during this period, concern about unemployment led to a focus on cutting the wage taxes employers pay to fund social security programs. Unemployment in the European Union (EU) shot past 9 percent in 1992, and has not fallen below that level since.[11] Studies suggest that wage taxes have contributed to unemployment in the EU by making workers more expensive to employers.[12] Thus wage tax cuts, which were absent in the first shifts in Sweden and Denmark, propelled almost every shift thereafter. Indeed, more than environmental worries, it is the presence of a major, tax-exacerbated social problem that has made tax shifting a political winner in Western Europe.

The year 1999 marked the surge of a second wave of tax shifting, this one spreading to some of Europe's economic heavyweights: Italy, France, and Germany, as well as the United Kingdom, which is preparing its second shift.[13]

In Germany, a historic "red-green" coalition of the Social Democrat and Green parties had recently gained power. The Greens conditioned their participation in the coalition on commitment to environmental tax reform, among other things.[14] They had hoped for a broad-based energy tax, but after much debate and political compromise, the parliament adopted a plan that exempted coal and jet fuel, and charged resource-intensive industries such as steel and forestry only 20 percent of the regular tax rate.[15] Altogether, the plan shifted some 2.1 percent of the government revenue base.[16]

In sum, the shifts in Western Europe have been small, and compromises have often been made in the fine print. Worldwide, moreover, environmental taxes generate barely 3 percent of all tax revenue, mainly through motor fuel levies.[17] Yet studies suggest that energy taxes will need to climb far above current levels to address problems such as global climate change—enough to generate perhaps 15 percent of all government revenue.[18] If taxes do eventually shift that far—and worldwide—then the small shifts of the past decade in Western Europe will be seen as the halting beginnings of a sweeping historical trend.

Satellites Boost Environmental Knowledge   Molly O. Sheehan

In 1997, satellite images of fires in Southeast Asia helped explain why rainforests, which are supposed to be wet, were burning so quickly. Some images of Indonesia showed the fires often concentrated in areas approved for commercial land use, and often started in the morning, when land clearing began.[1] Combined with other information, the satellite data helped researchers to conclude that the forests were being systematically burned to make way for palm oil plantations.[2]

This is just one example of how satellites have augmented our ability to understand Earth systems.[3] Early milestones in systematically monitoring the environment from space include the first weather satellite in 1960 and the first land observation satellite in 1972, both launched by the United States.[4] In 1999, more than 45 Earth observation missions were operating, and more than 70 are planned during the next 15 years by civil space agencies and private companies.[5]

Satellites are unique in being able to collect detailed information about parts of Earth that are otherwise difficult to access—the far reaches of the atmosphere, the depths of the oceans, and the icy polar regions. Moreover, remote-sensing instruments aboard Earth-orbiting satellites can frequently record changes over large areas and long periods of time.

Satellite sensors can also open up various parts of the electromagnetic spectrum to human observation by recording heat and reflected energy invisible to the human eye. For instance, emissions in the near-infrared range can be used to assess the health of vegetation (because healthy green vegetation reflects most of the near-infrared radiation it receives), and thermal radiation can reveal fires that would otherwise be obscured by smoke.[6] Some sensors use radar, transmitting short bursts of microwave energy to Earth and recording the strength of the reflected energy that comes back. Microwaves can penetrate the atmosphere in all conditions, so radar can "see" in the dark and through haze, clouds, or smoke.

Meteorological satellites form the backbone of the most effective global environmental monitoring program to date: the World Weather Watch. Operated by the World Meteorological Organization, this network combines satellite observations with ground, sea, and air monitoring stations, telecommunication links, and computer analysis centers.[7] In recent years, optical sensors that collect data on sea surface temperature and radar sensors that estimate ocean height have proved useful in understanding and predicting El Niño events, which bring warmth and wetness to much of the west coasts of South and North America and drought to Southeast Asia, Australia, and parts of Africa.[8]

In the 1990s, researchers began to delve into satellite archives to study longer-term climate patterns. For instance, satellite images have helped reveal a lengthening growing season in northern latitudes and the breakup of major ice sheets.[9] Radar sensors have been used to construct topographical maps of the ocean bottom, which in turn provide better understanding of the ocean currents, tides, and temperatures that affect climate.[10] Not until recently, however, did space agencies begin to design satellite systems dedicated specifically to climate research. In 1999, the United States launched Terra, which carries five different sensors for recording climatic variables such as radiative energy fluxes, clouds, water vapor, snow cover, land use, and the biological productivity of oceans.[11] It is to be the first in a series of satellites that will create a consistent long-term data set.

International organizations and national governments can use remote imaging to give more teeth to environmental laws. One leading fishing nation, Peru, is monitoring its coastal waters to prevent the kind of heavy overfishing that has caused fisheries to collapse.[12] In Italy, the city of Ancona plans to buy satellite images to detect illegal waste dumps.[13]

Governments have launched most of the current global fleet of Earth observation satellites, but private companies are now beginning to enter the picture. (See Table 1.) One of the first was OrbImage, a U.S. company

TABLE 1: SELECTED SATELLITE SYSTEMS PRODUCING COMMERCIALLY AVAILABLE IMAGERY

SATELLITE	LAUNCH DATE	OWNER	SPATIAL RESOLUTION[1]
Landsat series	1972	NASA (U.S.	30–120 meters
Landsat-7	1999	space agency)	15–60 meters
Terra	1999	NASA	15 meters–22 kilometers
SPOT series	1986	CNES (French	10–30 meters
SPOT-4	1997	space agency)	
AVHRR	1979	NOAA (U.S. agency)	1.1 kilometer
IRS-1D	1997	Indian remote sensing agency	6 meters
Ikonos	1999	Space Imaging Corp.	1–4 meters
OrbView-2/SeaWiFS	1997	OrbImage Corp.	1 kilometer
OrbView-3	2000		1–8 meters
OrbView-4	2000–01		1–8 meters
QuickBird	2000	Earthwatch Corp.	1–4 meters
Radarsat-1	1995	Canadian space agency	8–100 meters

[1]The lower end of the range usually applies to panchromatic (black and white) images, whereas the higher end applies to multispectral (color) images.
SOURCE: Web sites of various space agencies and companies.

that launched a satellite called SeaWiFS in 1997. Originally designed to measure ocean color and temperature, this satellite has monitored fires in Indonesia, floods in China, and dust storms in the Sahara and Gobi Deserts.[14]

In September 1999, U.S.-based Space Imaging launched the first of a new generation of high-resolution satellites, which will produce the most detailed images that can be bought on the open market.[15] OrbImage and Earthwatch, another U.S. company, are planning similar systems.[16] Whereas one picture element—or "pixel"—in a SeaWiFS image corresponds to 1 square kilometer on the ground, in the newer systems a pixel corresponds to just 1 square meter.

Different tasks require different levels of detail. Whereas the wide coverage provided by lower-resolution satellites has proved useful in understanding large-scale natural features, very detailed imagery may be best able to reveal niche habitats important for protect-

ing biodiversity and constructions such as buildings, tanks, weapons, and refugee camps. The new high-resolution imagery may be a powerful tool for watchdog groups that monitor arms control agreements and government military activities.[17]

Ultimately, an educated global citizenry will be needed to make use of the flood of data being unleashed from satellites. As Ann Florini of the Carnegie Endowment writes, "With states, international organizations, and corporations all prodding one another to release ever more information, civil society can take that information, analyze and compile it, and disseminate it to networks of citizen groups and consumer organizations."[18] The newest generation of "eyes in the sky" can produce highly detailed images—but many brains on the ground will be needed to make sure that these are put to good use.

Corporate Mergers Skyrocket

Michael Renner

The trend toward corporate consolidation gained significant momentum in 1999, when the value of worldwide mergers and acquisitions reached a new record of $3.4 trillion (in 1998 dollars).[1] (See Figure 1.) This was an astonishing 40-percent increase from the previous record of $2.5 trillion—established only in 1998.[2] Since 1980, the annual value of mergers has risen 100-fold, reaching a cumulative $15 trillion.[3] In 1999, more than 32,000 deals were announced, triple the number of 10 years earlier and more than 30 times as many as in 1981.[4]

In a sign of globalization gathering pace, the dollar value of mergers across national boundaries rose even more strongly than that of all mergers, almost doubling to $1.1 trillion in 1999.[5] While cross-border mergers were typically below 20 percent of the value of all mergers in the early 1980s, today they represent 33 percent.[6] The number of cross-border deals valued at more than $1 billion rose from 35 in 1997 to 89 in 1998.[7]

Mergers are held to increase shareholder value and boost corporate efficiency. But evidence suggests that these expectations are not always fulfilled.[8] From a broader vantage point, there is concern that mergers and a proliferation of strategic partnerships among corporations are leading to a greater degree of market concentration in many industries, giving a few producers an undue amount of influence on the market.[9] Market power often also translates into political influence.

Large corporations have enormous influence on how billions of people work and live. In recent years, for instance, concern has risen about the push by biotech firms to manipulate the genetic makeup of food and plants, about media giants' control over the way we learn about global events, about highly mobile companies weakening labor's bargaining position, about civic culture coming under the sway of corporate advertisements and sponsorships, and about industry lobbyists influencing the outcome of elections and legislation. Ultimately, consolidation trends may threaten democratic norms, labor standards, human rights, and environmental quality.[10]

The current merger frenzy is such that yesterday's record-shattering deal looks almost quaint today. In 1998, the biggest announced acquisition—Exxon's purchase of Mobil—was valued at $86 billion.[11] In 1999, the telephone giant MCI Worldcom proposed to buy its rival Sprint for $108 billion.[12] Early 2000 saw the announcement of the two largest combinations ever—between America Online and Time-Warner for $165 billion, and between Britain's Vodafone Airtouch and Germany's Mannesmann for $183 billion.[13] The merger record of 1999 is on track to be surpassed in 2000: the total value of announced deals just 11 weeks into the year is $864 billion—closing in on the total for 1995.[14]

Recent deals have established new behemoths in such diverse parts of the economy as telecommunications, pharmaceuticals, oil, automobiles, and paper.[15] In 1999, one third of the worldwide merger value was concentrated in just three sectors. The telecommunications industry, with $569 billion, was by far the leader, followed by commercial banking

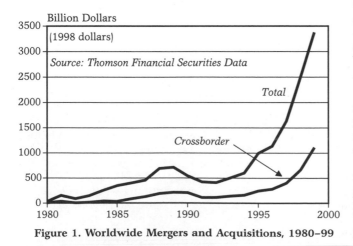

Figure 1. Worldwide Mergers and Acquisitions, 1980–99

($377 billion) and radio and television broad-casting ($246 billion).[16] Among cross-border deals, telecommunications, too, is the leader, followed by the metals, oil and gas, and chemical industries.[17]

Liberalization and privatization of telecom-munications assets in many countries have triggered an endless series of takeovers.[18] For example, if the announced merger between MCI Worldcom and Sprint is approved, it will be the culmination of 18 successive mergers over the past two decades, 11 of which were multibillion-dollar combinations.[19]

The media industry is being thoroughly reshaped by the growing integration of enter-tainment, news, publishing, and communica-tions companies and by the rapid rise of the Internet and digital communications.[20] Just nine corporate giants now dominate the world media market.[21]

Companies in Western Europe and North America are by far the most active in acquir-ing firms elsewhere; firms from all other regions of the world, including even Japan, are comparatively small players. Developing-country enterprises play a minuscule role as buyers, although they have been important takeover targets during the 1990s.[22]

Thomson Financial Securities Data reports that $487 billion worth of mergers, or 14 per-cent of the total in 1999, involved so-called hostile takeovers—offers that had initially been rejected by the target company.[23] At more than four times the value of hostile deals during the previous peak year, 1988, this marked yet another record for 1999. Corporate raiding had been rare outside the United States, but is now spreading else-where: more than one third of the dollar vol-ume of all European mergers in 1999 involved hostile acquisitions.[24]

Mergers in the 1980s were largely financed through "leveraged buy-outs"—borrowed money—and acquired firms were often canni-balized. Today, mergers involve stock swaps rather than cash transactions. Bloated stock market values have made deals of previously unimagined size possible.[25]

But the current cascade of mergers is also sustained by the broad trend toward privati-zation of state-owned companies and public infrastructure, deregulation, and the liberal-ization of trade, investments, and capital mar-kets.[26] In an age of globalization, the size and geographical reach of a firm are seen as ever more crucial to success. Increa-singly, firms either achieve this objective by absorbing others, or they get swallowed up by competitors.[27]

Cross-border mergers have been the main driving force of foreign direct investment (FDI) in recent years. This means that a con-siderable portion of private capital flows goes simply to changing ownership of existing fac-tories and other businesses.[28] While some acquisitions imply a long-term investment commitment, others may be little more than a prelude to asset-stripping—retaining the most valuable parts of a company and closing or selling off other parts.[29]

The tidal wave of cross-border mergers implies that transnational corporations (TNCs), particularly the largest ones, will con-tinue to expand their already strong role in world trade. Intra-firm trade—that is, the flow of raw materials, components, finished goods, and services from a subsidiary of a corpora-tion in one country to another subsidiary in a second country—now accounts for roughly a third of world trade.[30] The proportion rises to two thirds if what the World Investment Report 1999 calls "arm's-length trade associated with TNCs" is included.[31]

Through mergers and other FDI flows, transnational corporations can supply domes-tic markets in numerous countries through a growing web of local factories and offices. At $11 trillion in 1998, sales of foreign affiliates of transnationals easily surpassed total world exports of $6.7 trillion.[32] During the past decade, these sales have grown more strongly than either total world output or world trade.[33]

Social
Features

Wind Energy Jobs Rising

Michael Renner

Reducing fossil fuel use and shifting toward renewable energy sources such as wind and solar power is one of the key challenges in moving toward a sustainable economy. Opponents of such a shift have long argued that pursuing an alternative energy path would be a job killer.[1] But this need not be the case. Among alternative sources of energy, wind power has progressed most rapidly, and it is now beginning to offer a fast-increasing number of jobs.

Wind power development opens up employment opportunities in a variety of fields, including meteorologists and surveyors to select and rate appropriate sites, structural engineers to design the turbines and supervise their assembly, metal workers to supply the rotors, and mechanics and computer operators to monitor the system and keep it in good working order.

Numerous studies find that wind power compares favorably in its job-creating capacity with coal- and nuclear-generated electricity. In Germany, although wind energy contributed a still minuscule 1.2 percent of total electricity generation in 1998, it provided some 15,000 jobs in manufacturing, installing, and operating wind machines.[2] In comparison, nuclear power had 31 percent of the electricity market but supported a comparatively meager 40,000 jobs; coal-generated power had a 26-percent market share and was the source of 80,000 jobs.[3] Given the rapid expansion of wind power in Germany, wind will likely overtake nuclear as a source of jobs in 2000.[4]

Wind power generation is mostly decentralized and small-scale, and the manufacturing of rotor blades and other components requires skilled labor input to ensure quality. Although the increasing size of wind turbines and growing economies of scale will in coming years translate into somewhat fewer jobs relative to each unit of energy produced, wind will still compare favorably with traditional electricity sources.[5]

The lion's share of the world's wind power-generating capacity has been installed in Europe. And because European companies are the leading manufacturers of wind turbines, most of the world's wind power–related jobs are being generated there. A 1996 study found that some 16,000 jobs were created in the Danish wind power industry, a world leader.[6] The European Wind Energy Association (EWEA, an industry group) projects that up to 40 gigawatts of wind power capacity could be installed in Europe by 2010, creating between 190,000 and 320,000 jobs.[7]

Wind power is now poised to move from a marginal source of energy to a major contributor in many parts of the world. *Windforce 10*, a study released in October 1999 by EWEA, Greenpeace International, and the Forum for Energy and Development, contends that wind energy could meet 10 percent of the world's electricity demand by 2020.[8] The report assessed the number of jobs that might be generated under this scenario. It relied on the results of the most comprehensive national study undertaken to date, done by the Danish Wind Turbine Manufacturers Association. This assumes that 17 job-years of employment are created for every megawatt of wind energy capacity manufactured and an additional 5 job-years for the installation of every megawatt, for a total of 22 job-years.[9] *Windforce 10* accounts for rising labor productivity, estimating that the per-megawatt job figures will gradually decrease to 15.5 by 2010 and 12.3 by 2020.[10]

On the basis of these assumptions, the study projected worldwide wind power employment to increase from about 57,000 jobs in 1998 and 67,000 jobs in 1999 to 1.7 million over the next two decades.[11] (See Figure 1.) In fact, this may be an underestimate. Because new installations during 1999 surpassed the study's assumptions, the number of jobs supported that year was likely even higher—perhaps on the order of 86,000.[12]

Offshore wind installations are expected to play a growing role in coming years, particularly in Europe. These will require larger investments and support greater employment—

a development not reflected in Figure 1. These job numbers also do not include employment generated through additional investments, as required, to enlarge nations' electrical infrastructure.[13]

The study's jobs-per-megawatt formula appears to be well within the range of other reports. The European Commission, for example, noted in a 1997 report that, as a rough rule of thumb, 1 megawatt of wind power generating capacity installed creates jobs for 15–19 people under present European market conditions, and perhaps double that in countries with lower labor productivity.[14] In a 1997 study, Greenpeace Germany estimated that 14 jobs are created by manufacturing and installing 1 megawatt.[15]

Additional employment is generated through operating and maintaining wind turbines. EWEA reports that in Europe, between 100 and 450 people are employed per year for every terawatt-hour of electricity produced, depending on the age and type of turbine used.[16] In 1999, that would have meant anywhere from 3,000 to 13,000 additional jobs. As wind power expands, so will these numbers.[17]

European companies accounted for about 90 percent of worldwide wind turbine sales in 1997, and presumably will continue to garner the majority of jobs in the near future.[18] As other regions with high wind power potential gear up, they will only realize job gains if they master the technology. India and China, for instance, can in principle generate substantial wind power employment if they succeed in strengthening their indigenous production base. India already has 14 domestic turbine manufacturers, and spare parts production and turbine maintenance are helping at least some regions and villages generate much-needed income and employment.[19] Argentina hopes to create 15,000 permanent

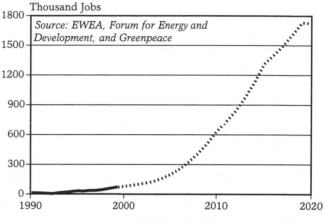

Thousand Jobs

Source: EWEA, Forum for Energy and Development, and Greenpeace

Figure 1. World Wind Power Employment, 1990–99, with Projections to 2020

jobs over the next decade.[20]

Asian, Latin American, and East European countries currently have labor productivity rates in the wind power sector that are estimated to be 20 percent lower than in Western Europe.[21] This means that domestically manufactured wind turbines create one fifth more jobs than those imported from Western Europe. However, Asian countries will likely continue to rely on imports for some 20 percent of their installations during the next decade; Latin American and East European nations are principally able to manufacture all needed components within their own regions.[22] The Middle East and Africa, by contrast, will mostly depend on imported technology and components.[23]

Tuberculosis Resurging Worldwide   Lisa Mastny

In the last few decades, tuberculosis (TB) has experienced a dramatic resurgence worldwide. Some 1.8 billion people—nearly one third of the world—now carry the TB bacteria, *Mycobacterium tuberculosis*.[1] The World Health Organization (WHO) predicts that if control is not strengthened, nearly 1 billion more will be infected by 2020—many of whom will die from TB.[2]

This age-old disease is spread when an individual with active pulmonary TB coughs or sneezes, releasing tiny infectious droplets into the air.[3] Once inhaled, the bacteria can lie dormant in someone's body for years—even a lifetime.[4] It attacks when the immune system is weak, and affects most tissues and organs, but particularly the lungs.[5] Each sick person infects on average 10–15 new people a year, although only about 5–10 percent of the individuals carrying TB actually come down with the disease.[6]

TB kills nearly 2 million people a year and is the leading cause of death among women of reproductive age.[7] In the developing world, an estimated 60 percent of TB cases occur in the most productive age group, ages 15 to 44.[8] Families must pay not only the direct costs of TB prevention and treatment, but also the indirect costs of lost labor time from the illness.[9]

Tuberculosis was a leading cause of death in northern Europe and the Americas until about 1900.[10] During much of the twentieth century, improved living conditions and new anti-TB drugs and other effective treatments helped vanquish the disease in most industrial nations.[11] But in the mid-1980s, TB again surged in many of these countries, including the United States.[12] The disease has spread rapidly in the former Soviet Union and Eastern Europe, where the number of reported cases jumped by more than 25 percent between 1994 and 1996 alone.[13]

Yet the developing world continues to bear the brunt of the epidemic. Of the estimated 8 million new TB cases reported worldwide in 1998, roughly 95 percent were in developing countries—with 3 million in Southeast Asia (two thirds of them in India alone); 2 million in the western Pacific region, which includes China; and 1.6 million in Africa.[14] (See Table 1.) These regions also register the most TB deaths.

In recent years, the single largest factor behind the TB surge has been the spread of AIDS, particularly in Asia and Africa.[15] Because AIDS weakens the immune system, an HIV-positive person is up to 30 times likelier than someone else to develop TB.[16] In 1998, an estimated 20 percent of the world's TB victims were HIV-positive.[17] This combination has been deadliest in Africa, where nearly 60 percent of TB victims also had HIV.[18]

Another factor behind the rise in TB has been the emergence of new drug-resistant strains, which are harder to control and up to 100 times costlier to treat.[19] These strains can emerge when patients cut short their treatment or take the wrong mix of drugs, enabling the bacteria to develop resistance, which is then passed on.[20] An estimated 50 million people may now be infected with drug-resistant TB, with the highest rates in the Baltic States, Russia, the Dominican Republic, and Côte d'Ivoire.[21]

The boom in international travel and tourism, as well as increased migration, have also contributed to the spread of TB.[22] Airplanes provide ideal environments for transmission, as they often are crowded and poorly ventilated.[23] On one Paris–New York flight in 1998, a Ukrainian man sick with TB may have infected 13 of the 40 passengers sitting nearest to him, although the direct transfer could not be verified.[24]

TB has spread quickly among the world's rising refugee and immigrant populations, who are highly mobile and often face conditions of malnutrition and overcrowding.[25] As they move from place to place, they may carry the disease with them—in many industrial countries, foreign-born persons account for at least half of all TB cases.[26]

TB is also rampant among the world's 10–30 million prisoners, who show TB rates 5–20 times higher than national averages.[27] An estimated half of Russia's million prisoners

TABLE 1: TUBERCULOSIS INCIDENCE AND DEATHS, AND CO-INFECTION WITH HIV, 1998

REGION	INCIDENCE			DEATHS		
	TOTAL INCIDENCE (thousand new cases)	SHARE OF WORLD TOTAL (percent)	SHARE HIV-POSITIVE (percent)	TOTAL DEATHS (thousand)	SHARE OF WORLD TOTAL (percent)	SHARE HIV-POSITIVE (percent)
Southeast Asia	3,002	37	2	717	38	5
Western Pacific	1,984	25	< 1	360	19	1
Africa	1,557	19	33	514	28	59
Eastern Mediterranean	611	8	1	142	8	2
Europe	39	5	2	64	3	6
Americas	421	5	6	68	4	21
World	8,012	100	8	1,863	100	20

SOURCE: World Health Organization, *World Health Report 1999* (Geneva: 1999). Figures are preliminary estimates.

are infected with TB and one tenth have the active disease, often a drug-resistant strain.[28] Prison rooms are typically unclean, over-crowded, and poorly ventilated, and prisoners face high rates of malnutrition, drug use, and HIV infection.[29] Rapid prison turnover also means that an estimated four to six times the prison population passes through a facility each year, increasing the chance of infecting new arrivals as well as people on the outside.[30]

Fortunately, TB is both preventable and treatable.[31] The BCG vaccine, invented in 1921, now has 85-percent global coverage and can help prevent certain types of TB in infancy.[32] But it has proved less effective in stemming TB in adulthood, particularly in AIDS-prone areas, and scientists are seeking to develop a more comprehensive alternative.[33]

One of the most cost-effective strategies for combating TB is DOTS: Directly Observed Treatment, Short-course.[34] Adopted by WHO and its partners on a global scale in 1994, DOTS involves treating patients with up to four drugs for six to eight weeks, for as little as $11 per person.[35] Since 1990, more than a million people have been treated this way.[36] The number of countries using the strategy

has risen 10-fold to more than 100—including 20 of the 22 most TB-afflicted countries.[37] DOTS has a consistent cure rate of about 85 percent, and in some countries, such as Peru, Bangladesh, and parts of China, treatment success has neared 95 percent.[38]

Still, only some 15 percent of the world's TB patients receive DOTS, according to WHO.[39] Impediments to implementation include a lack of political commitment or funding by governments, political turmoil and war, erratic drug supplies, and deteriorating public health care systems.[40] In Russia and Eastern Europe, only some 5 percent of TB patients are treated with DOTS, largely because of insufficient health care funding.[41]

But DOTS alone may not be enough to stem the global TB onslaught—particularly in countries with high rates of HIV or drug-resistant TB. In these areas, tougher control strategies will be needed that allow for more rigorous treatment and that address underlying social issues such as health care and poverty.[42]

Prison Populations Exploding Gary Gardner

Some 8 million people were held in prisons and jails in the mid- to late 1990s in the 180 countries for which data are available.[1] The total reflects inmates who are serving sentences as well as those awaiting trial. It does not, however, include millions more who are denied their freedom in non-prison circumstances such as forced labor. When these are included, the global prison population could be 10–30 million.[2]

In at least 18 countries, prisoners held in pre-trial detention constitute the majority of the prison population, according to the international monitoring group Human Rights Watch.[3] In some countries, pre-trial detainees spend more time locked up than if they had been tried and convicted.

Just three countries—the United States, China, and Russia—hold more than half of the global prison population.[4] The United States has 1.7 million persons behind bars, while China has 1.4 million and Russia has 1 million.[5] (See Table 1.) On a per capita basis, Russia and the United States had by far the largest prison populations in the late 1990s: 687 and 645 people per 100,000 population, respectively.[6] The global average is 137 prisoners per 100,000 population.[7]

National rates of imprisonment reflect particular mixes of criminal, law enforcement, judicial, and penal characteristics. Japan's exceptionally low rate of incarceration—just 39 people per 100,000 population—is consistent with its low rate of crime.[8] But in other countries a low rate might mean that police, courts, or corrections systems are not well developed; these factors might allow many criminals to avoid jail time.[9] Similarly, high imprisonment rates can reflect the work of a repressive police state, a high rate of crime, long prison sentences, or the use of mandatory sentences.

The top three incarcerating countries use jails heavily for different reasons. Russia has a long history of imprisonment: one of every four males in Russia has spent some time in jail, according to the Moscow Center for Prison Reform.[10] Most are held in work camps rather than in western-style prisons. The camps once manufactured a variety of products for sale to the public, generating great profit for the government. Today, however, the market for prison manufactures is limited and the camps are experiencing acute financial difficulties.[11]

China's rate of imprisonment appears to be on the low side, at 113 people per 100,000. But official figures may represent only 13 percent of those deprived of freedom.[12] Dissident Harry Wu, a former prisoner who has studied

TABLE 1: HIGHEST AND LOWEST PRISON POPULATIONS AND RATES OF IMPRISONMENT IN COUNTRIES WITH AT LEAST 20 MILLION POPULATION, LATE 1990s

NUMBER OF PRISONERS		IMPRISONMENT RATE (per 100,000 population)	
HIGHEST			
United States	1,700,000	Russia	687
China	1,410,000	United States	645
Russia	1,010,000	Ukraine	413
India	231,325	South Africa	321
Ukraine	211,568	Uzbekistan	258
LOWEST			
Nepal	6,200	Indonesia	20
Peru	20,899	India	24
Uganda	21,971	Nepal	29
Malaysia	24,400	Bangladesh	37
Venezuela	25,000	Japan	39

SOURCE: Roy Walmsley, "World Population Prison List," Research Findings No. 88 (London: Development and Statistics Directorate, Home Office Research, 1999).

the Chinese system, estimates that 4–6 million people are sentenced to "reform through labor," 3–5 million are in "re-education" labor camps, and 8–10 million are forced to work in prison factories or farms.[13] Millions more are held in pre-trial detention.[14]

In the United States, the prison population has risen rapidly since the 1970s, when state and federal governments began to require mandatory and increasingly lengthy prison sentences for drug possession.[15] The population in state and federal prisons grew from fewer than 200,000 inmates in 1970 to 1.2 million in 1998, with another 600,000 in local jails.[16] Some 36 percent of prisoners entering state prisons and 71 percent of those in federal prisons were convicted of drug offenses.[17] The drug-driven rapid increase in prison populations has led to widespread overcrowding; California's system, for example, is running at twice its intended capacity—despite the construction of 21 new prisons in the past 20 years.[18]

Overcrowding is common in prison systems globally as well. Combined with poor sanitation and lack of food and health care, overcrowding facilitates the spread of disease, including killers such as tuberculosis and AIDS. In many countries, tuberculosis cases run 5–20 times higher in prison than in society at large; in Russia, the prevalence is 40 times higher in prison than outside.[19] AIDS, too, is increasing rapidly among some prison populations. In the United States, new cases appear in prison at five times the frequency found in the general population.[20]

The past decade has seen a shift toward privately run prisons in some countries. Although these hold less than 2 percent of the world's prisoners, the concept is spreading rapidly: the number of beds in private prisons globally increased more than ninefold between 1990 and 1999.[21] Some 85 percent of these—158 facilities—are found in the United States.[22] Australia, England, Netherlands Antilles, New Zealand, Scotland, and South Africa have another 30 private prisons between them.[23]

Private prisons are promoted as a less expensive way to handle incarceration, since they are often cheaper to build and operate than state-run facilities.[24] But critics charge that the savings come at the expense of other considerations, including just wages for guards, and health care and programs for prisoners.[25] And because private prisons, like hotels, are most profitable when run at peak capacity, they create an incentive to maximize incarceration.[26]

For many, the debate on public versus private prisons begs the question of whether prison is the answer to most crime. In many countries, drug offenses are handled through treatment programs rather than through imprisonment. Arizona recently adopted such an approach. Because imprisonment costs the state $50 per day, while treatment, counseling, and probation run just $16 per day, Arizona saved more than $2.5 million the first year of the change in policy.[27] More than three quarters of the people on probation stayed free of drugs thus far.[28]

Even for non-drug offenses, prisons may do more harm than good if their primary purpose is to punish rather than to rehabilitate. "Prison is an expensive way of making bad people worse," noted David Waddington, U.K. Home Secretary in 1989–91.[29] Often, prison does not address the underlying causes of crime. More than 70 percent of prison inmates in the United States had a history of drug abuse before entering prison, but only 10 percent received drug treatment once inside.[30] And 70 percent are reported to be illiterate.[31] With these handicaps, it is little wonder that two thirds of U.S. inmates are rearrested within three years of release.[32]

Women Slowly Gain Ground in Politics   Molly O. Sheehan

As of December 1999, women accounted for 13.2 percent of the representatives in the lower or popular chambers of national legislatures worldwide, according to the Geneva-based Inter-Parliamentary Union.[1] Women have slowly been gaining ground in this arena since World War II (see Table 1), with the proportion of women in the lower chambers of parliaments nearly quadrupling from 3 percent in 1945 to 11.6 percent in 1995.[2]

Although women constitute more than half of world population and play an important role in environmentally sustainable development, until well into the twentieth century many nations denied women the right to vote and run for office.[3] New Zealand in 1893 and Australia in 1902 were the first to grant electoral rights to women but these laws applied only to women of European descent; today, only Bahrain, Kuwait, and the United Arab Emirates continue to bar women from full political participation.[4] In Kuwait, the Emir introduced a measure in 1999 to allow women to vote and run in elections, but the nation's all-male Parliament rejected the plan.[5] Although most countries now allow women to vote and stand for election, there is a long way to go to achieve equal political participation.

Some regions of the world boast a greater percentage of women politicians than others. A worldwide ranking of the number of women members in the lower chamber of national parliaments taken every 10 years between 1945 and 1995 found Finland and Sweden consistently in the top 15.[6] As a group, the Nordic countries have the highest share of women in national legislatures. (See Table 2.)

Women have attained the top posts in national governments only recently. Of the 35 women heads of state in the twentieth century, 28 were elected in the 1990s.[7] In June 1999, lawmakers in Latvia made Vaira Vike-Freiberga the first democratically elected woman president in Eastern Europe.[8] In Indonesia, Megawati Sukarnoputri was the front-runner in the nation's first democratic election, in 1999.[9] Although she won a majority of the popular vote, she could not summon enough electoral votes in the parliament, which eventually elected her Vice President.[10]

In a few countries, the profile of women has risen in the latest round of elections. For instance, a record number of women candidates for parliament and local councils were on the ballots in Turkey's

TABLE 1: WOMEN IN NATIONAL LEGISLATURES, SELECTED COUNTRIES, 1955–99[1]

COUNTRY	1955	1975	1985	1999
		(percent)		
Sweden	12.2	21.4	31.5	42.7
Germany[2]	8.8/24.5	5.8/31.8	9.8/32.4	30.9
South Africa	1.9	0.6	1.1	30.0
Cuba	—	—	22.6	27.6
Namibia	—	—	—	22.2
China	12.0	22.6	21.2	21.8
Costa Rica	6.7	5.3	7.0	19.3
United Kingdom	3.8	3.6	3.5	18.4
Mexico	2.5	6.0	11.0	18.2
United States	2.7	3.7	5.0	13.3
France	3.7	2.7	7.1	10.9
Soviet Union/Russia	24.0	30.5	31.1	10.2
India	4.0	4.1	7.8	8.4
Brazil	0.3	0.3	1.5	5.7
Japan	1.7	1.4	1.6	4.6

[1]Percent who are members of single-chamber parliaments or in the lower chamber of two-chamber systems. Dashes indicate that elections had not yet been held. [2]Figures for 1955, 1975, and 1985 are for West/East Germany.
SOURCE: IPU, *Women in Parliaments 1945–1995: A World Statistical Survey* (Geneva: 1995); IPU, "Women in National Parliaments: Situation as of 5 December 1999," < www.ipu.org >.

TABLE 2: WOMEN IN NATIONAL
LEGISLATURES, REGIONAL AVERAGES,
1999[1]

REGION	SHARE (percent)
Nordic countries	38.9
Americas	14.7
Asia	14.6
Rest of Europe	12.3
Sub-Saharan Africa	11.2
Pacific	10.9
Arab States	3.6

[1]In countries with bicameral systems, includes both houses.
SOURCE: IPU, "Women in National Parliaments: Situation as of 5 December 1999," <www.ipu.org>.

1999 elections.[11] Voters in Mexico City in 1999 elected a woman mayor, considered the second most powerful post in the nation after the presidency; during the same year, a woman was elected president of a national political party there, the Party of the Democratic Revolution, for the first time.[12] And although the 46 women running for election to the Northern Ireland Assembly in 1998 represented only 15 percent of the field, it was still a greater share than ever before.[13]

Some gains have also been made at the level of national ministers. Italy now has twice as many women Cabinet ministers as in any previous government.[14] After the 1999 election in South Africa, women constituted 38 percent of the 42-person cabinet.[15]

Barriers to female participation in politics vary from country to country. A recent study in the United States found that newspapers cover women candidates for executive-branch positions in state and national government differently from their male counterparts. Coverage of women tends to focus more on the candidates' personal characteristics and less on their positions on issues.[16]

Some countries have used quotas to boost the number of women in power. In India, following a 1993 amendment reserving one third

of all seats in local elections for women, more than 800,000 women were elected.[17] Similarly, a surge of women candidates entered Brazil's local elections in 1996 after a law required that at least 20 percent of each political party's candidates be women.[18] Other countries with some form of quota system include Argentina, Finland, Germany, Mexico, South Africa, and Spain.[19]

Various groups have arisen to promote female leadership.[20] For example, Sisterhood is Global, a network of more than 1,300 individuals and organizations in 70 countries, educates women about their rights.[21] In many Islamic countries, such groups are helping to create a momentum for change.[22] A Council of Women World Leaders, established at Harvard University's Kennedy School of Government in 1996, draws on the experience of current and former women heads of state to encourage women to participate in politics.[23]

A number of international decisions have helped legitimize the political involvement of women. Relevant treaties include the 1952 Convention on the Political Rights of Women and the 1979 Convention on the Elimination of Discrimination Against Women. In 1995, the United Nations sponsored the fourth world conference on women in Beijing. With 189 governments and 2,600 nongovernmental groups in attendance, this was one of the largest U.N. conferences ever.[24] Delegates agreed to a set of strategic objectives and actions, including efforts to advance the role of women in politics and environmental stewardship.[25] A special session of the United Nations in New York in June 2000 will assess progress toward these goals.

NOTES

GRAIN HARVEST FALLS
(pages 34–35)

1. U.S. Department of Agriculture (USDA), Foreign Agricultural Service (FAS), *Grain: World Markets and Trade* (Washington, DC: February 2000).
2. Ibid.; USDA, *Production, Supply, and Distribution*, electronic database, Washington, DC, updated February 2000.
3. USDA, op. cit. note 2; U.S. Bureau of the Census, *International Data Base*, electronic database, Suitland, MD, updated 28 December 1999.
4. USDA, op. cit. note 1.
5. USDA, op. cit. note 1; USDA, op. cit. note 2.
6. USDA, op. cit. note 1.
7. Ibid.
8. Ibid.
9. USDA, op. cit. note 2.
10. Ibid.
11. Ibid.
12. Ibid.
13. Ibid.
14. Ibid.
15. Ibid.
16. Ibid.
17. Ibid. Days of consumption are calculated by dividing annual global grain consumption by 365 and then dividing the result by world carryover stocks.
18. Ibid.
19. Projected 3-percent rise in global economy from International Monetary Fund, *World Economic Outlook* (Washington, DC: October 1999).

SOYBEAN HARVEST DROPS
(pages 36–37)

1. U.S. Department of Agriculture (USDA), Foreign Agricultural Service (FAS), *Oilseeds: World Markets and Trade* (Washington, DC: February 2000); USDA, *Production, Supply, and Distribution (PS&D)*, electronic database, Washington, DC, updated February 2000.
2. USDA, *Oilseeds*, op. cit. note 1; U.S. Bureau of the Census, *International Data Base*, electronic database, Suitland, MD, updated 28 December 1999.
3. USDA, *Oilseeds*, op. cit. note 1.
4. USDA, *PS&D*, op. cit. note 1.
5. Ibid.
6. Ibid.
7. USDA, *Oilseeds*, op. cit. note 1.
8. Ibid.
9. Ibid.
10. Ibid.
11. Ibid.
12. Ibid.
13. Ibid.
14. USDA, *PS&D*, op. cit. note 1.
15. Ibid.
16. USDA, *Oilseeds*, op. cit. note 1.
17. Ibid.
18. Ibid.
19. Ibid.
20. Ibid.
21. Ibid.
22. One tenth as food from ibid.

MEAT PRODUCTION UP AGAIN
(pages 38–39)

1. U.N. Food and Agriculture Organization (FAO), *FAOSTAT Statistics Database*, < apps.fao.org >, updated 7 December 1999.
2. Ibid.
3. U.S. Department of Agriculture (USDA), Foreign Agricultural Service (FAS), *Livestock and Poultry: World Markets and Trade* (Washington, DC: October 1999).
4. Ibid.
5. Ibid.
6. Ibid.
7. Ibid; trend since 1990 from FAO, op. cit. note 1.
8. FAO, op. cit. note 1.

9. USDA, op. cit. note 3.
10. Ibid.
11. Ibid.
12. FAO, op. cit. note 1.
13. USDA, op. cit. note 3.
14. Ibid.
15. Ibid.
16. FAO, op. cit. note 1.
17. USDA, op. cit. note 3.
18. Ibid.
19. Ibid.
20. Ibid.
21. FAO, op. cit. note 1.
22. FAO, op. cit. note 1; U.S. Bureau of the Census, *International Data Base*, electronic database, Suitland, MD, updated 28 December 1999.
23. FAO, op. cit. note 1.
24. USDA, op. cit. note 3.
25. Ibid.
26. Ibid.
27. Ibid.
28. Ibid.

FISH HARVEST DOWN (pages 40–41)

1. Catch in 1998 from U.N. Food and Agriculture Organization (FAO), Fisheries Web site, <www.fao.org/fi/statist/summtab/default.asp>, viewed 27 January 2000; 1997 catch from FAO, *Yearbook of Fishery Statistics: Capture Production 1997*, vol. 84 (Rome: 1999).
2. FAO, *Aquaculture Production Statistics 1988–1997* (Rome: 1999).
3. Worldwatch calculation based on data in FAO, Fisheries Web site, op. cit. note 1, in FAO, *Yearbook*, op. cit. note 1, and in FAO, op. cit. note 2.
4. Worldwatch calculation based on data in FAO, Fisheries Web site, op. cit. note 1, and on FAO, op. cit. note 2; U.S. Bureau of the Census, *International Data Base*, electronic database, Suitland, MD, updated 28 December 1999.
5. National Marine Fisheries Service, *Fisheries of the United States, 1998* (Washington, DC: U.S. Department of Commerce, July 1999).
6. FAO, Fisheries Web site, op. cit. note 1.
7. Ibid.
8. FAO, *Review of the State of the World Fishery Resources: Marine Fisheries* (Rome: 1997).
9. Worldwatch calculation based on data in FAO, *Yearbook of Fishery Statistics: Catches and Landings* (Rome: various years); 1990–97 data from FAO, *Yearbook*, op. cit. note 1.
10. Richard Grainger, "Recent Trends in Global Fishery Production," FAO, <www.fao.org/fi/trends/catch/catch.asp>, viewed 28 January 2000.
11. Daniel Pauly et al., "Fishing Down Aquatic Food Webs," *American Scientist*, January/February 2000.
12. FAO, op. cit. note 9.
13. Roland Schmitten, National Marine Fisheries Service, discussion with author, 21 January 2000.
14. Worldwatch calculation based on data in FAO Fisheries Web site, op. cit. note 1, in FAO, *Yearbook*, op. cit. note 1, and in FAO, op. cit. note 2.
15. FAO, Fisheries Web site, op. cit. note 1.
16. Rosamond L. Naylor et al., "Nature's Subsidies to Shrimp and Salmon Farming," *Science*, 30 October 1998.
17. Ibid.
18. Ibid.
19. Ibid.
20. FAO, <www.fao.org/fi/highligh/2010.asp>, viewed 25 January 2000.
21. Worldwatch calculation based on fish projections in FAO, op. cit. note 20, and on global population from Bureau of Census, op. cit. note 4.
22. Ibid.

GRAIN AREA SHRINKS AGAIN (pages 44–45)

1. U.S. Department of Agriculture (USDA), *Production, Supply, and Distribution*, electronic database, Washington, DC, February 2000.
2. Area per person calculated from global grain area from USDA, op. cit. note 1; population from U.S. Bureau of the Census, *International Data Base,* electronic database, Suitland, MD, updated 28 December 1999.
3. Tim Dyson, *Population and Food: Global Trends and Future Prospects* (London: Routledge, 1996).
4. USDA, *Grain: World Markets and Trade* (Washington, DC: December 1999).
5. Ibid.
6. Vaclav Smil, "China's Agricultural Land," *The China Quarterly*, June 1999.
7. USDA, Farm Service Agency, <www.fsa.usda.gov/pas/news/releases/1999/10/0391.htm>, viewed 17 January 2000.
8. Smil, op. cit. note 6.
9. USDA, Natural Resources Conservation Service, "1997 National Resources Inventory: Highlights," <www.nhq.nrcs.usda.gov/land/pubs/97highlights.html>, viewed 9 January 2000.

10. Ibid.
11. Ibid.
12. Ibid.
13. Smil, op. cit. note 6.
14. USDA, op. cit. note 9.
15. USDA, op. cit. note 1.
16. Worldwatch estimate based on grain area from USDA, op. cit. note 1, and population from Bureau of Census, op. cit. note 2. Estimate assumes continuation of a long-standing historical trend: nations whose grain area per person falls below 0.06–0.07 hectares per person import at least 20 percent of their grain.
17. Niels H. Batjes, "Management Options for Reducing CO_2-Concentrations in the Atmosphere by Increasing Soil Sequestration in the Soil," ISRIC Technical Paper 30 (Wageningen, the Netherlands: International Soil Reference and Information Centre, 1999).

FERTILIZER USE DOWN
(pages 46–47)

1. K.G. Soh and K.F. Isherwood, "Short Term Prospects for World Agriculture and Fertilizer Use," presentation at 25th IFA Enlarged Council Meeting, International Fertilizer Industry Association (IFA), Rome, 30 November–3 December 1999; additional data provided by Claudine Aholou-Putz, IFA, e-mail to Brian Halweil, Worldwatch Institute, 27 January 2000.
2. Soh and Isherwood, op. cit. note 1; Aholou-Putz, op. cit. note 1.
3. Soh and Isherwood, op. cit. note 1; Aholou-Putz, op. cit. note 1; U.S. Bureau of the Census, International Data Base, electronic database, Suitland, MD, updated 28 December 1999.
4. Soh and Isherwood, op. cit. note 1; Aholou-Putz, op. cit. note 1; U.N. Food and Agriculture Organization (FAO), Fertilizer Yearbook (Rome: various years); Bureau of Census, op. cit. note 3.
5. Soh and Isherwood, op. cit. note 1.
6. Ibid.
7. Ibid.
8. Ibid.
9. Ibid.
10. Ibid.; Aholou-Putz, op. cit. note 1; FAO, op. cit. note 4.
11. Soh and Isherwood, op. cit. note 1; Aholou-Putz, op. cit. note 1.
12. Soh and Isherwood, op. cit. note 1.
13. Ibid.; Aholou-Putz, op. cit. note 1; FAO, op. cit. note 4.
14. Soh and Isherwood, op. cit. note 1; Aholou-Putz, op. cit. note 1.
15. Soh and Isherwood, op. cit. note 1; Aholou-Putz, op. cit. note 1.
16. Soh and Isherwood, op. cit. note 1; Aholou-Putz, op. cit. note 1.
17. Soh and Isherwood, op. cit. note 1.
18. Sandra Postel, Pillar of Sand (New York: W.W. Norton & Company, 1999).
19. Paul Epstein et al., Marine Ecosystems: Emerging Diseases as Indicators of Change, Health, Ecological and Economic Dimensions of Global Change Program (Boston: Center for Health and Global Environment, Harvard Medical School, December 1998).

PESTICIDE TRADE NEARS NEW HIGH (pages 48–49)

1. U.N. Food and Agriculture Organization (FAO), FAOSTAT Statistics Database, < apps.fao.org >, viewed 17 December 1999. World totals for exports and imports are not equal in the figures given by FAOSTAT because of differences in reporting on trade by individual countries. Both total imports and total exports do, however, follow generally the same trends over time.
2. FAO, op. cit. note 1.
3. Arnold L. Aspelin and Arthur H. Grube, Pesticide Industry Sales and Usage: 1996 and 1997 Market Estimates (Washington, DC: Environmental Protection Agency (EPA), 1999); Helmut F. van Emden and David B. Peakall, Beyond Silent Spring: Integrated Pest Management and Chemical Safety (London: Chapman & Hall, 1996).
4. Barbara Dinham, The Pesticide Hazard: A Global Health and Environmental Audit (London: Zed Books, 1993); Pesticide Action Network North America (PANNA), "Growth in 1994 World Agrochemical Market," Global Pesticide Campaigner, June 1995; The Pesticides Trust, "Review of the Global Pesticide Market," Pesticide News, December 1993.
5. FAO, op. cit. note 1.
6. Ibid.
7. Ibid.
8. Ibid.
9. Ibid.
10. Ibid.
11. Ibid.
12. Ibid.; PANNA, "World and US Agrochemical Market in 1998," PANUPS, 23 July 1999.
13. PANNA, "World Pesticide Market Expands,"

PANUPS, 29 August 1997.

14. Dinham, op. cit. note 4; van Emden and Peakall, op. cit. note 3; The Pesticides Trust, "The Pesticide Business—Impact on Food Security," *Pesticides News*, December 1998; Montague Yudelman et al., "Pest Management and Food Production: Looking to the Future," Food, Agriculture, and the Environment Discussion Paper 25 (Washington, DC: International Food Policy Research Institute, September 1998).

15. World pesticide use estimate from David Pimental, Cornell University, Ithaca, NY, e-mail to author, 23 September 1999; FAO, op. cit. note 1; Aspelin and Grube, op. cit. note 3.

16. It is important to note that in industrial countries, farmers have been increasingly using pesticides with lower dosages. Additionally, a much higher percentage of industrial-nation farmers than developing-country farmers already use pesticides. Thus growth in pesticide intensity in industrial regions has slowed.

17. The Pesticides Trust, op. cit. note 4; The Pesticides Trust, op. cit. note 14.

18. Arnold L. Aspelin, *Pesticide Industry Sales and Usage: 1994 and 1995 Market Estimates* (Washington, DC: EPA, 1997).

19. PANNA, "Pesticides Updates—Asia," *PANUPS*, 16 October 1998; PANNA, op. cit. note 4; Aspelin, op. cit. note 18. Use of conventional pesticides in the United States fell 4 percent from 438.6 million kilograms of active ingredients in 1992 to 421.4 million kilograms in 1993. Almost three quarters of this decline was due to a reduction in pesticide use in agriculture, mainly because of flooding in Midwestern states. Some 13.2 million fewer kilograms of pesticide were used in agriculture in 1993 than in 1992.

20. In 1986, Sweden decided to reduce agricultural pesticide use by 50 percent; by 1991, the nation had achieved a 47-percent reduction and committed to achieving another 50-percent reduction; David Pimental, *Techniques For Reducing Pesticide Use: Economic and Environmental Benefits* (West Sussex, UK: J. Wiley & Sons, 1997). Denmark also inaugurated a 50-percent pesticide reduction plan in 1986, and it is looking to shift to 100-percent organic agriculture within the next decade; The Pesticides Trust, "Denmark—Action to Reduce Use," *Pesticides News*, September 1998.

21. FAO, op. cit. note 1; Yudelman et al., op. cit. note 14; van Emden and Peakall, op. cit. note 3.

22. FAO, "Rotterdam Convention on Harmful Chemicals and Pesticides Adopted and Signed," press release (Rome: 11 September 1998).

23. FAO, "Ratifications, Signatories and Conference Participants of the Convention and the Final Act (As of 1 December 1999)," list of signatories of the Rotterdam Convention, <www.fao.org/ag/agp/agpp/pesticid/pic/convlist.htm>, viewed 19 January 2000.

FOSSIL FUEL USE IN FLUX
(pages 52–53)

1. Figure for 1999 is a preliminary Worldwatch estimate based on BP Amoco, *BP Amoco Statistical Review of World Energy 1999* (London: Group Media & Publications, June 1999), on U.S. Department of Energy (DOE), Energy Information Administration (EIA), *Monthly Energy Review January 2000* (Washington, DC: January 2000), on DOE, EIA, *International Energy Outlook 1999* (Washington, DC: March 1999), on PlanEcon, Inc., *PlanEcon Energy Outlook* (Washington, DC: October 1999), on European Commission (EC), *Energy in Europe: 1999 Annual Energy Review*, Special Issue (Brussels: January 2000), on Eurogas, "Natural Gas Consumption in Western Europe Continued its Growth in 1999" (Brussels: February 2000), on International Monetary Fund (IMF), *World Economic Outlook* (Washington, DC: October 1999), and on Jonathan E. Sinton and David G. Fridley, "What Goes Up: Recent Trends in China's Energy Consumption" (Berkeley, CA: Lawrence Berkeley National Laboratory, 9 December 1999).

2. Numbers for 1950–98 based on United Nations, *World Energy Supplies 1950–74* (New York: 1976), on British Petroleum (BP), *BP Statistical Review of World Energy* (London: Group Media & Publications, various years), and on DOE, EIA, *International Energy Database*, May 1998.

3. Worldwatch estimate based on DOE, *Monthly Energy Review*, op. cit. note 1, on EC, op. cit. note 1, on PlanEcon, op. cit. note 1, and on IMF, op. cit. note 1.

4. Worldwatch estimate based on IMF, op. cit. note 1.

5. Worldwatch estimate based on DOE, *Monthly Energy Review*, op. cit. note 1.

6. Worldwatch estimate based on ibid., on DOE, *International Energy Outlook*, op. cit. note 1, on EC, op. cit. note 1, and on PlanEcon, op. cit. note 1.

7. EC, op. cit. note 1; DOE, *International Energy Outlook*, op. cit. note 1.

8. Worldwatch estimate based on BP Amoco, op. cit. note 1; Worldwatch categorizes the use of natural gas liquids as natural gas consumption, whereas other organizations include these numbers under oil consumption. This difference explains why other datasets may show coal use still exceeding that of natural gas.

9. Worldwatch estimate based on BP Amoco, op. cit. note 1, on DOE, *Monthly Energy Review*, op. cit. note 1, on EC, op. cit. note 1, on PlanEcon, op. cit. note 1, and on Sinton and Fridley, op. cit. note 1.

10. Ibid.; BP, op. cit. note 2.

11. Sinton and Fridley, op. cit. note 1; PlanEcon, op. cit. note 1; EC, op. cit. note 1.

12. Andrew Taylor, "Moves to Save Coal Industry Misfired, Says Study," *Financial Times*, 19 January 2000.

13. Ibid.

14. National Acid Precipitation Assessment Program, *NAPAP Biennial Report to Congress: An Integrated Assessment* (Silver Spring, MD: May 1998); J.L. Stoddard et al., "Regional Trends in Aquatic Recovery from Acidification in North America and Europe," *Nature*, 7 October 1999; Andrew C. Revkin, "New York To Sue 17 Power Plants on Air Pollution," *New York Times*, 15 September 1999; Matthew L. Wald, "Old Plants With New Parts Present a Problem to E.P.A.," *New York Times*, 26 December 1999.

15. Sinton and Fridley, op. cit. note 1.

16. James Kynge, "China Starts to Pull the Plug on Smoky Power Plants," *Financial Times*, 13 January 2000.

17. Ibid.

18. Oil prices based on BP Amoco, op. cit. note 1, on BP, op. cit. note 2, and on DOE, *Monthly Energy Review*, op. cit. note 1.

19. Curtis Rist, "Why We'll Never Run Out of Oil," *Discover*, June 1999.

20. Amy Myers Jaffe and Robert A. Manning, "The Shocks of a World of Cheap Oil," *Foreign Affairs*, January/February 2000.

21. Rist, op. cit. note 19.

22. C.J. Campbell, *The Coming Oil Crisis* (Essex, UK: Multi-Science Pub. Co. & Petroconsultants, 1997).

23. Rist, op. cit. note 19; Mark Hertsgaard, "Will We Run Out of Gas?" *Time*, 8 November 1999.

NUCLEAR POWER RISES SLIGHTLY (pages 54–55)

1. Installed nuclear capacity is defined as reactors connected to the grid as of 31 December 1999, and is based on Worldwatch Institute database complied from statistics from the International Atomic Energy Agency (IAEA) and press reports primarily from *Associated Press, Reuters, Agence Press de France, Financial Times*, and World Wide Web sites.

2. Worldwatch Institute database, op. cit. note 1.

3. Ibid.

4. Ibid.

5. Ibid.

6. Ibid. Excluded from the list of operating reactors are seven reactors in Canada that are officially "laid up" for repairs. Although not listed as permanently closed yet, it appears highly unlikely that three of these reactors will ever operate again. The other four may be brought back on-line in the next few years. Data for 1993 are from IAEA, *Nuclear Power Reactors in the World* (Vienna: 1994).

7. Darrel Nash, *1999 Nuclear Plant Vulnerability Study* (Arlington, VA: Energy Access LLC, October 1999).

8. U.S. Department of Energy, Energy Information Administration, *Annual Energy Outlook 2000, with Projections to 2020* (Washington, DC: December 1999).

9. *Nuclear Notes from France*, No. 62, October–November 1999, < info-france-use.org/america/embassy/nuclear/n2f2/octobe99.htm >, viewed 14 January 2000.

10. "20 Years After Landmark Vote, Sweden Shuts Down a Nuclear Reactor," *Agence Press de France*, 30 November 1999; "Nuclear Plant in Britain to Close," *Associated Press*, 1 December 1999; Uranium Institute, "Netherlands," News Briefing, 8–14 December 1999.

11. "German Energy Industry to Shut Down Four Atom Plants by 2002," *Agence Press de France*, 6 December 1999.

12. "Bulgaria to Close N-Plants," *Financial Times*, 1 December 1999.

13. Jonathan Leff, "Lithuanian N-Plant Casts Shadow over EU," *Reuters*, 7 December 1999; "World's First Nuclear Power Plant Plans Closure," *Reuters*, 10 June 1999; "Nuclear Cloud Over New Wave of EU Enlargement," *Agence Press de France*, 5 December 1999.

14. Randy Fabi, "US-Kazakhstan Agree to Close Nuclear Reactor," *Reuters*, 21 December 1999.

15. Scott Stoddard, "Plant Worker Dies of Radiation," *Daily Camera* (Boulder, CO), 22 December 1999.

16. Baku Nishio, "Plan to Build 20 New Reactors— Unlikely to Succeed," Citizens Nuclear Information Center, < www.jca.ax.apc.org/

cnic/english/topics/20readtors.html>, viewed 14 January 2000.

17. "China Puts a Hold on New Nuclear Units," *The Electricity Daily*, 7 May 1999.

18. "S. Korea Plans Major Electricity Capacity Expansion," *Reuters*, 13 January 2000.

19. "India Aims for 20,000 MWs of Nuclear Power by 2020," *Reuters*, 30 September 1999.

20. "Turkey to Start Building Nuclear Plant in Early 2000," *Agence Press de France*, 11 December 1999.

21. Turkey, *Uranium Institute News Briefing*, Uranium Institute, 22 December 1999/4 January 2000, <www.uilondon.org/nb/nb00/nb0001.htm>, viewed 21 January 2000.

WIND POWER BOOMS (pages 56–57)

1. Worldwatch Institute preliminary estimate based on figures from European Wind Energy Association (EWEA), "Europe Beats Year 2000 Wind Energy Target," press release (London: 24 January 2000), on American Wind Energy Association (AWEA), *Global Wind Energy Market Report*, January 2000, at <www.awea.org>, on "Germany: Spectacular Growth in 1999," *Renewable Energy Report*, February 2000, on Andreas Wagner, Fordergesellschaft Windenergie (FGW), Hamburg, Germany, e-mail to author, 10 February 2000, and on Jose Santamarta, Madrid, Spain, e-mail to author, 19 February 2000.

2. Historical series from BTM Consult, *International Wind Energy Development: World Market Update 1998* (Ringkobing, Denmark: March 1999).

3. Worldwatch estimate based on sources cited in note 1; "Mobile Phone Sales are Taking Wing," *Washington Post*, 2 March 2000.

4. Worldwatch value estimate based on $800 per kilowatt turbine cost; for job estimate, see "Wind Energy Jobs Rising" in this volume.

5. "Germany: Spectacular Growth in 1999," op. cit. note 1.

6. Andreas Wagner, "Building on Success—Germany's New Renewable Energy Law" (Hamburg, Germany: FGW, February 2000).

7. Jose Santamarta, independent analyst, Madrid, Spain, e-mail to author, 19 February 2000.

8. "Spain: EHN Orders 1,800 Wind Turbines," *Renewable Energy Report*, February 2000.

9. Historical series from BTM Consult, op. cit. note 2.

10. Christine Real de Azua, "The U.S. Wind Market

is Back," *Solar Today*, January 2000.

11. Paul Gipe, Paul Gipe & Associates, Tehachapi, CA, discussion with author, 8 February 2000.

12. AWEA, op. cit. note 1.

13. EWEA, op. cit. note 1.

14. *Windpower Monthly*, various issues.

15. David Milborrow, "Wind Narrows the Price Gap Again," *Windpower Monthly*, January 2000.

16. Ibid.

17. Jos Beurskens, "Going to Sea: Wind Goes Offshore," *Renewable Energy World*, January 2000.

18. Ibid.

19. EWEA, Forum for Energy and Development, and Greenpeace International, *Wind Force 10: A Blueprint to Achieve 10% of the World's Electricity from Wind Power by 2020* (London: 1999).

20. Ibid.

SOLAR POWER MARKET JUMPS
(pages 58–59)

1. Paul Maycock, "1999 World Cell/Module Production," *PV News*, March 2000, and discussion with author, 12 February 2000.

2. Growth estimate based on Paul Maycock, *PV News*, various issues.

3. Market estimate based on average module cost of $3.50 per watt and average system cost of $8 per watt.

4. Maycock, op. cit. note 1.

5. Ibid.

6. Ibid.

7. Ibid.

8. Ibid.

9. "Million Solar Roofs Cites Progress," *PV News*, May 1999.

10. Matthew L. Wald, "Where Some See Rusting Factories, Government Sees a Source of Solar Energy," *New York Times*, 4 August 1999.

11. Maycock, op. cit. note 1.

12. Ibid.

13. Ibid.

14. "Solar Cell Manufacturing Capacity Expanding Fast," *Renewable Energy World*, January 2000.

15. "PV Technology Lightens Burden of Village Women," *PV News*, January 2000.

16. Rene Karottki and Douglas Banks, "PV Power and Profit? Electrifying Rural South Africa," *Renewable Energy World*, January 2000.

17. Maycock, op. cit. note 1.

18. Ibid.

19. Ibid.

20. Ibid.

21. "Renewable Energy Tracker: January 2000," *Renewable Energy Report*, February 2000.

COMPACT FLUORESCENTS LIGHT UP THE GLOBE (pages 60–61)

1. Nils Borg, International Association for Energy Efficient Lighting (IAEEL), Stockholm, e-mail to author, 14 January 2000.
2. Nils Borg, IAEEL, Stockholm, e-mail to author, 26 January 2000.
3. Ibid.
4. Data for 1988–89 from Evan Mills, Lawrence Berkeley Laboratory, Berkeley, CA, letter to Worldwatch, 3 February 1993; 1990–99 estimate from Borg, op. cit. note 1.
5. Worldwatch estimate based on a 15-percent decay in existing compact fluorescent lamp (CFL) stock per year and the same amount of light from 15-watt CFLs and 60-watt incandescents.
6. Worldwatch estimate based on 15-watt CFLs replacing 60-watt incandescents, used for four hours per day (17 percent availability), and where an "average-sized coal fired power plant" is a 440-megawatt electric output, operating 80 percent of the time (80 percent availability).
7. Global incandescent sales are estimated at 11 billion units in 1999; assumes life of 1,000 hours for an incandescent lamp and 10,000 hours for a CFL. Incandescent-market estimate received from Nils Borg, IAEEL, Stockholm, e-mail to author, 17 January 2000. The 1999 CFL market sales figure (432 million) is decayed by 15 percent over the year.
8. Worldwatch estimate based on 15-percent decay in existing CFL stock per year, 15-watt CFLs replacing 60-watt incandescents, four hours of lighting per day, and 0.196 tons of carbon and 0.0038 tons of sulfur dioxide per 1,000 kilowatt-hours of electricity generated in the United States. While the estimate expresses the savings from installing CFLs instead of incandescents, it is difficult to determine how many CFLs are literally used in place of incandescents. Electricity consumption and carbon emissions from U.S. Department of Energy, Energy Information Administration, "1997 U.S. Electric Information at a Glance," <www.eia.doe.gov>, viewed 19 January 2000.
9. Worldwatch estimate based on a 10,000-hour 15-watt CFL costing 4 Euro replacing a 1,000-hour 60-watt incandescent costing 0.5 Euro; lamps used four hours per day, electricity costs 0.13 Euro per kilowatt hour; Casper Kofod, Lighting Consultant, Denmark, e-mail to author, 29 December 1999.
10. Worldwatch estimate based on a 5-percent discount rate in the net present value calculation; energy and bulb prices from Kofod, op. cit. note 9.
11. Peter DuPont, International Institute for Energy Conservation—Asia office, Bangkok, e-mail to author, 13 January 2000.
12. China 1996 and 1999 sales data estimates from Borg, op. cit. note 1.
13. Asian CFL consumption and sales data combine China, Japan, and other Asian states; estimate from Borg, op. cit. note 1.
14. "Green Light for Europe", *IAEEL Newsletter*, January 1999.
15. Russell Sturm, International Finance Corporation, e-mail to author, 20 January 2000.
16. Ibid.
17. Barry Bredenkamp, Eskom, Johannesburg, South Africa, e-mail to author, 17 January 2000.
18. Ibid.
19. Robert Henderson, Eskom, Johannesburg, South Africa, e-mail to author, 25 January 2000.
20. Steve Johnson, Lawrence Berkeley Laboratory, e-mail to author, 14 January 2000.

GLOBAL TEMPERATURE DROPS (pages 64–65)

1. James Hansen et al., NASA Goddard Institute for Space Studies, Surface Temperature Analysis, "Global Land-Ocean Temperature Index in .01 C," at <www.giss.nasa.gov/data/update/gis temp>, viewed 11 January 2000.
2. Ibid.
3. National Oceanic and Atmospheric Administration (NOAA), National Climatic Data Center, "Climate of 1999: Annual Review—Preliminary Report," 13 December 1999; World Meteorological Organization (WMO), "1999 Closes the Warmest Decade and Warmest Century of the Last Millennium According to WMO Annual Statement on the Global Climate," press release (Geneva: 16 December 1999).
4. M.E. Mann, R.S. Bradley, and M.K. Hughes, "Northern Hemisphere Temperatures During the Past Millennium: Inferences, Uncertainties, and Limitations," *Geophysical Research Letters*, 15 March 1999.
5. NOAA, op. cit. note 3.
6. Ibid.
7. Ibid.

8. Ibid.
9. Ibid.
10. Ibid.
11. Ibid.
12. Ibid.
13. J. Hansen et al., "GISS Analysis of Surface Temperature Change," *Journal of Geophysical Research*, December 1999; Hansen et al., op. cit. note 1.
14. Michael J. McPhaden, "The Child Prodigy of 1997–98," *Nature*, 15 April 1999.
15. M. Lockwood, R. Stamper, and M.N. Wild, "A Doubling of the Sun's Coronal Magnetic Field During the Past 100 Years," *Nature*, 3 June 1999; E.N. Parker, "Sunny Side of Global Warming," *Nature*, 3 June 1999; Simon F.B. Tett, "Causes of Twentieth-Century Temperature Change Near the Earth's Surface," *Nature*, 10 June 1999.
16. Martin I. Hoffert et al., "Solar Variability and the Earth's Climate," *Nature*, 21 October 1999; Tett, op. cit. note 15.
17. National Research Council, Board on Atmospheric Sciences and Climate, *Reconciling Observations of Global Temperature Change* (Washington, DC: National Academy Press, January 2000).
18. Ibid.
19. Tom M.L. Wigley, *The Science of Climate Change: Global and U.S. Perspectives* (Arlington, VA: Pew Center on Global Climate Change, June 1999).
20. Ibid.
21. Ibid.

CARBON EMISSIONS FALL AGAIN
(pages 66–67)

1. Figure for 1999 is a preliminary Worldwatch estimate based on BP Amoco, *BP Amoco Statistical Review of World Energy 1999* (London: Group Media & Publications, June 1999), on U.S. Department of Energy (DOE), Energy Information Administration (EIA), *Monthly Energy Review January 2000* (Washington, DC: 2000), on DOE, EIA, *International Energy Outlook 1999* (Washington, DC: March 1999), on PlanEcon, Inc., *PlanEcon Energy Outlook* (Washington, DC: October 1999), on European Commission (EC), *Energy in Europe: 1999 Annual Energy Review*, Special Issue (Brussels: January 2000), on Eurogas, "Natural Gas Consumption in Western Europe Continued its Growth in 1999" (Brussels: February 2000), on International Monetary Fund (IMF), *World Economic Outlook* (Washington, DC: October 1999), and on Jonathan E. Sinton and David G. Fridley, "What Goes Up: Recent Trends in China's Energy Consumption" (Berkeley, CA: Lawrence Berkeley National Laboratory, 9 December 1999); figures for 1950–98 are from emissions data (including gas flaring, but excluding cement production) from G. Marland et al., "Global, Regional, and National Fossil Fuel CO_2 Emissions," in *Trends: A Compendium of Data on Global Change* (Oak Ridge, TN: Carbon Dioxide Information Analysis Center, Oak Ridge National Laboratory, DOE, March 1999).
2. Worldwatch estimate based on Marland et al., op. cit. note 1, and on BP Amoco, op. cit. note 1.
3. Worldwatch estimate based on Marland et al., op. cit. note 1, and on BP Amoco, op. cit. note 1; Worldwatch update of Angus Maddison, *Monitoring the World Economy, 1820–1992* (Paris: Organisation for Economic Co-operation and Development (OECD), 1995), using deflators and recent growth rates from IMF, op. cit. note 1.
4. Worldwatch estimate based on Marland et al., op. cit. note 1, and on BP Amoco, op. cit. note 1.
5. Worldwatch estimate based on Marland et al., op. cit. note 1, and on BP Amoco, op. cit. note 1.
6. International Energy Agency (IEA), *CO_2 Emissions from Fuel Combustion, 1971–1996: 1998 Edition* (Paris: OECD/IEA, 1998).
7. Worldwatch estimate based on Marland et al., op. cit. note 1, and on BP Amoco, op. cit. note 1.
8. Worldwatch estimate based on Marland et al., op. cit. note 1, on BP Amoco, op. cit. note 1, and on DOE, EIA, *Emissions of Greenhouse Gases in the United States 1998* (Washington, DC: October 1999).
9. Worldwatch estimate based on Marland et al., op. cit. note 1, and on BP Amoco, op. cit. note 1; IEA, op. cit. note 6.
10. Worldwatch estimate based on Marland et al., op. cit. note 1, and on BP Amoco, op. cit. note 1.
11. Worldwatch estimate based on Marland et al., op. cit. note 1, and on BP Amoco, op. cit. note 1.
12. IEA, op. cit. note 6; Worldwatch estimate based on Marland et al., op. cit. note 1, on BP Amoco, op. cit. note 1, and on Worldwatch update of Maddison, op. cit. note 3.
13. William K. Stevens, "Global Economy Slowly Cuts Use of High-Carbon Energy," *New York Times*, 31 October 1999.
14. Worldwatch estimate based on Marland et al., op. cit. note 1, on BP Amoco, op. cit. note 1,

and on Worldwatch update of Maddison, op. cit. note 3; Robert M. Margolis and Daniel M. Kammen, "Underinvestment: The Energy Technology and R&D Policy Challenge," *Science*, 30 July 1999.

15. "Better News—Perhaps—On Global Warming" (editorial), *Nature*, 5 August 1999; Tony Reichhardt, "Emissions Fall Despite Economic Growth," *Nature*, 5 August 1999; John J. Fialka, "Flat CO_2 Emissions Give Experts Hope," *Wall Street Journal*, 2 August 1999; Stevens, op. cit. note 13.

16. Earth Negotiations Bulletin (ENB), "Summary of the Fifth Conference to the Parties to the Framework Convention on Climate Change, 25 October–5 November 1999" (Winnipeg, MN, Canada: International Institute for Sustainable Development, 8 November 1999).

17. Sebastian Oberthur and Hermann Ott, *The Kyoto Protocol: International Climate Policy for the 21st Century* (New York: Springer-Verlag, 1999).

18. Ibid.; Michael Grubb with Christiaan Vrolijk and Duncan Brack, *The Kyoto Protocol: A Guide and Assessment* (London: Royal Institute of International Affairs and Earthscan Publications, 1999).

19. Jae Edmonds et al., *International Emissions Trading and Global Climate Change: Impacts on the Costs of Greenhouse Gas Mitigation* (Arlington, VA: Pew Center on Global Climate Change, December 1999); ENB, op. cit. note 16.

20. ENB, op. cit. note 16.

21. C.D. Keeling and T. Whorf, Scripps Institution of Oceanography, La Jolla, CA, e-mail to author, 1 February 2000; C.D. Keeling and T. Whorf, "Atmospheric CO_2 Records from Sites in the SIO Air Sampling Network," in *Trends*, op. cit. note 1.

22. Keeling and Whorf, e-mail, op. cit. note 21; Keeling and Whorf, "Atmospheric CO_2 Records," op. cit. note 21.

23. Richard A. Feely, "Influence of El Niño on the Equatorial Pacific Contribution to Atmospheric CO_2 Accumulation," *Nature*, 15 April 1999; "1998 Global CO_2 Emissions Decline, While Concentrations Jump," *Global Environmental Change Report*, 13 August 1999.

24. J.R. Petit et al., "Climate and Atmospheric History of the Past 420,000 Years from the Vostok Ice Core, Antarctica," *Nature*, 3 June 1999.

25. Ibid.

ECONOMIC GROWTH SPEEDS UP
(pages 70–71)

1. Worldwatch updates of Angus Maddison, *Monitoring the World Economy 1820–1992* (Paris: Organisation for Economic Co-operation and Development (OECD), 1995), using deflators and recent growth rates from International Monetary Fund (IMF), *World Economic Outlook* (Washington, DC: October 1999).

2. IMF, op. cit. note 1.

3. Ibid.

4. Ibid.

5. Ibid.

6. Ibid.

7. Ibid.

8. Ibid.

9. Ibid.

10. Ibid.

11. Ibid.; poverty figures are from UNICEF, cited in World Bank, *East Asia Regional Overview* (Washington, DC: September 1999), and show the change from mid-1997 to mid-1998.

12. OECD, *National Accounts* (Paris: various years).

13. Herman E. Daly and John B. Cobb, Jr., *For the Common Good* (Boston: Beacon Press, 1989); Marilyn Waring, *If Women Counted: A New Feminist Economics* (San Francisco, CA: HarperSanFrancisco, 1991); U.N. Development Programme (UNDP), *Human Development Report 1995* (New York: Oxford University Press, 1995).

14. On air pollution, see Gene M. Grossman and Alan B. Krueger, "Economic Growth and the Environment," *Quarterly Journal of Economics*, May 1995.

15. Mark Anielski and Jonathan Rowe, *The Genuine Progress Indicator: 1998 Update* (San Francisco, CA: Redefining Progress, 1999).

16. Ibid.

17. Ibid.

18. Worldwatch updates of Maddison, op. cit. note 1, using deflators and recent growth rates from IMF, op. cit. note 1, and population data from U.S. Bureau of the Census, *International Data Base*, electronic database, Suitland, MD, updated 28 December 1999.

19. Robert Costanza et al., "The Value of the World's Ecosystem Services and Natural Capital," *Nature*, 15 May 1997. Figure converted to 1998 dollars.

20. UNDP, op. cit. note 13. Figure converted to 1998 dollars.

DEVELOPING-COUNTRY DEBT INCREASES (pages 72–73)

1. World Bank, *Global Development Finance 1999*, electronic database, Washington, DC, 1999.
2. World Bank, *Global Development Finance 1999* (Washington, DC: 1999).
3. Ibid.
4. Ibid.
5. Ibid.; short-term debt has an original maturity of one year or less.
6. World Bank, op. cit. note 2.
7. Ibid.
8. World Bank, op. cit. note 1.
9. Judith Mann, "Missing the Big Picture on Debt Relief," *Washington Post*, 18 August 1999; Jubilee 2000 Coalition, "Who's Dropped What," <jubilee2000uk.org/reports/dropped.html>, viewed 1 February 2000.
10. Dean Murphy, "Debt Relief Has Two Sides," *Los Angeles Times*, 27 January 2000.
11. World Bank, op. cit. note 1.
12. Ibid.
13. The World Bank classifies a country's indebtedness on the basis of two ratios: total debt service to exports and total debt service to gross national product. In the case of the 40 countries designated as Highly Indebted Poor Countries, debt sustainability is also measured according to the ratio of debt to exports. A debt-to-exports ratio above 200–250 percent is considered unsustainable. World Bank, op. cit. note 1.
14. Jubilee 2000 Coalition, "Who We Are," <www.jubilee2000uk.org/main.html>, viewed 30 January 2000.
15. Jubilee 2000 Coalition, op. cit. note 9.
16. Ibid.
17. Jubilee 2000 Coalition, "UK Chancellor Acts on Debt," <oneworld.org/jubilee2000/gbrown.htm>, viewed 18 September 1998; "Gold Comfort," *The Economist*, 2 October 1999.
18. World Bank, *Global Development Finance 1998* (Washington, DC: 1998); Hilary French, "Coping with Ecological Globalization," in Lester R. Brown et al., *State of the World 2000* (New York: W.W. Norton & Company, 2000).
19. Jubilee 2000 Coalition, op. cit. note 9.

WORLD TRADE STABLE IN VALUE (pages 74–75)

1. Goods exports from International Monetary Fund (IMF), *International Financial Statistics,* electronic database, Washington, DC, March 2000; service exports from Barbara d'Andrea-Adrian, World Trade Organization (WTO), Geneva, e-mail to author, 16 February 2000; figures for 1999 are Worldwatch estimates, based on partial quarterly and monthly data from IMF, op. cit. this note. The goods and services exports data are computed on somewhat different bases and are not perfectly comparable. See technical notes to WTO, *International Trade Statistics* (Geneva: 1998).
2. Worldwatch estimates, op. cit. note 1.
3. IMF, op. cit. note 1; d'Andrea-Adrian, op. cit. note 1.
4. Price drop is a Worldwatch estimate using inflation-adjusted dollars, based on the Export Unit Value Index from IMF, op. cit. note 1.
5. IMF, op. cit. note 1; U.S. GNP Implicit Price Deflator from *Survey of Current Business,* various issues.
6. IMF, op. cit. note 1.
7. Ibid. Data for Germany include only West Germany before 1990.
8. Ibid.
9. Ibid.; Worldwatch estimates, op. cit. note 1. Data for China exclude Hong Kong.
10. IMF, *Balance of Payment Statistics Yearbook*, Part 2 (Washington, DC: 1999).
11. Ibid.
12. IMF, *Balance of Payment Statistics Yearbook*, Part 1 (Washington, DC: 1999).
13. Ibid.
14. IMF, op. cit. note 1; d'Andrea-Adrian, op. cit. note 1; Angus Maddison, *Monitoring the World Economy, 1820–1992* (Paris: Organisation for Economic Co-operation and Development, 1995), with recent growth rates from IMF, *World Economic Outlook* (Washington, DC: October 1999).
15. Biplab Dasgupta, *Structural Adjustment, Global Trade and the New Political Economy of Development* (London: Zed Books, 1998).
16. Ibid.
17. Ibid.
18. Ibid.
19. Hilary French, "Coping with Ecological Globalization," in Lester R. Brown et al., *State of the World 2000* (New York: W.W. Norton & Company, 2000).
20. Ibid.

WEATHER DAMAGES DROP (pages 76–77)

1. Figure for 1999 from Angelika Wirtz, Geoscience Research Group, Munich Re, e-mail to author, 3 February 2000. All Munich Re data

are strictly claims-related. Monetary values adjusted to 1998 dollars.

2. Data for 1980–98 from Munich Re, MRNatCatService, electronic database, Munich, Germany, 1998.

3. Ibid.

4. Figure for 1999 from Wirtz, op. cit. note 1; 1998 number from Munich Re, "Weather-Related Natural Disasters 1998" (Munich, Germany: 9 February 1999).

5. Munich Re, op. cit. note 2.

6. Munich Re, MRNatCatService, "Significant Natural Disasters in 1999," Munich, Germany, February 1999. As of 3 February 2000, loss evaluations were not yet finished for the Indian cyclone, Venezuelan floods and landslides, and European winter storms.

7. Raymond Colitt and Dan Bilefsky, "Cold Comfort for the Flood's Survivors," *Financial Times*, 22 December 1999; Munich Re, op. cit. note 6.

8. Munich Re, op. cit. note 6.

9. Suzanne Daley, "22 Die as France is Hit by 2nd Part of Weather's One-Two," *New York Times*, 29 December 1999.

10. Ibid.

11. Neelesh Misra, "Aid Reaches Cyclone Victims," *Washington Post*, 2 November 1999; Fred Pearce, "An Unnatural Disaster," *New Scientist*, 6 November 1999.

12. Pearce, op. cit. note 11; Robert Marquand, "Cyclone Damage Blocks Aid Effort," *Christian Science Monitor*, 2 November 1999; Barry Bearak, "Cyclone May Have Killed Thousands on India's East Coast," *New York Times*, 1 November 1999.

13. Bearak, op. cit. note 12.

14. Dennis S. Mileti, *Disasters by Design: A Reassessment of Natural Hazards in the United States* (Washington, DC: National Academy of Sciences, 1999).

15. Ibid.; Guy Gugliotta, "Cost of Natural Disasters Growing," *Washington Post*, 20 May 1999.

16. David L. Marcus and David Whitman, "More Sound Than Fury," *U.S. News and World Report*, 27 September 1999; Sharon Begley and Thomas Hayden, "Floyd's Watery Wrath," *Newsweek*, 27 September 1999.

17. Kenneth E. Kunkel, Roger A. Pielke, Jr., and Stanley A. Changnon, "Temporal Fluctuations in Weather and Climate Extremes That Cause Economic and Human Health Impacts: A Review," *Bulletin of the American Meteorological Society*, June 1999.

18. James P. Bruce, Ian Burton, and I.D. Mark Egener, *Disaster Mitigation and Preparedness in a Changing Climate: A Synthesis Paper* (Toronto: Emergency Preparedness Canada, 1999).

19. International Federation of Red Cross and Red Crescent Societies, *World Disasters Report 1999* (Geneva: 1999).

20. Ibid.; Frances Williams, "'Super-Disasters' May Be On Way," *Financial Times*, 25 June 1999.

PAPER PILES UP (pages 78–79)

1. U.N. Food and Agriculture Organization (FAO), *FAOSTAT Statistics Database*, < apps.fao.org >, viewed 26 January 2000.

2. Figure for 1950 from International Institute for Environment and Development (IIED), *Towards a Sustainable Paper Cycle* (London: 1996); population from U.S. Bureau of the Census, *International Data Base*, electronic database, Suitland, MD, updated 28 December 1999.

3. Shushuai Zhu, David Tomberlin, and Joseph Buongiorno, *Global Forest Products Consumption, Production, Trade and Prices: Global Forest Products Model Projections to 2010* (Rome: Forest Policy Planning Division, FAO, December 1998).

4. Miller Freeman, Inc., *International Fact and Price Book 1999* (San Francisco: 1998).

5. Ibid.

6. FAO, op. cit. note 1; population from Bureau of Census, op. cit. note 2; industrial countries are defined by FAO as "developed" (industrialized nations plus countries in transition).

7. U.S. consumption from Miller Freeman, Inc., op. cit. note 4; industrial figures from FAO, op. cit. note 1.

8. World from Miller Freeman, Inc., op. cit. note 4; developing countries from FAO, op. cit. note 1.

9. FAO, op. cit. note 1.

10. Figure of 29 percent from ibid.; 45 percent of total value from Bruce Michie, Cherukat Chandrasekharan, and Philip Wardle, "Production and Trade in Forest Goods," in Matti Palo and Jussi Uusivuori, eds., *World Forests, Society and Environment* (Dordrecht, Netherlands: Kluwer Academic Publishers, 1999).

11. Paul Hawken, Amory Lovins, and Hunter Lovins, *Natural Capitalism* (New York: Little Brown, 1999).

12. Miller Freeman, Inc., op. cit. note 4.

13. Ibid.

14. Ibid.

15. FAO, op. cit. note 1.

16. Wood Resources International Ltd., in *Fiber*

Sourcing Analysis for the Global Pulp and Paper Industry (London: IIED, September 1996), reported that in 1993 wood pulp production required approximately 618 million cubic meters of wood—equal to 18.9 percent of the world's total wood harvest in 1993 and 41.8 percent of the total industrial wood harvest (wood harvest volumes as noted by FAO, op. cit. note 1).

17. Wood Resources International Ltd., op. cit. note 16.

18. Paper growing faster than other major wood products from FAO, *State of the World's Forests 1999* (Rome: 1999).

19. World Commission on Forests and Sustainable Development, *Our Forests Our Future* (Cambridge, U.K.: Cambridge University Press, 1999).

20. IIED, op. cit. note 2.

21. Ibid.

22. Ibid.

23. Janet N. Abramovitz and Ashley T. Mattoon, *Paper Cuts: Recovering the Paper Landscape*, Worldwatch Paper 149 (Washington, DC: Worldwatch Institute, December 1999).

24. IIED, op. cit. note 2.

25. U.S. Environmental Protection Agency (EPA), *1997 Toxics Release Inventory Public Data Release Report* (Washington, DC: 1997).

26. Municipal solid waste burden in United States from Franklin Associates, Ltd., "Characterization of Municipal Solid Waste in the United States: 1998 Update," report prepared for EPA, Municipal and Industrial Solid Waste Division, Office of Solid Waste, 1999; European countries from IIED, op. cit. note 2.

GOLD LOSES ITS LUSTER
(pages 80–81)

1. Gold price is in 1998 dollars, based on historical data from Kitco Precious Metals Inc., online database, <www.kitco.com/charts/historical gold.html>, viewed 11 January 2000.

2. Data supplied by Earle Amey, Commodity Specialist, U.S. Geological Survey (USGS), letter to author, 10 January 2000, and discussion with author, 10 February 2000.

3. Ibid.

4. Ibid.

5. Lehman Brothers, Inc., *Reverse Alchemy: The Commoditization of Gold Accelerates* (New York: January 2000), based on data from the International Monetary Fund (IMF); Amey, discussion with author, op. cit. note 2.

6. Lehman Brothers, op. cit. note 5.

7. Ibid.

8. Ibid.

9. Gold prices are London market afternoon fix averages, from Kitco Precious Metals, op. cit. note 1; U.S. gross national product price deflator from U.S. Commerce Department, *Survey of Current Business*, various issues.

10. Dutch sales from Lehman Brothers, op. cit. note 5; Central Bank sales from Gold Field Mineral Services, Ltd. (GFMS), "Publication of Annual Gold Survey–*Gold 1998*," press release (London: 14 May 1998); GFMS, "Publication of *Gold Survey 1998–Update 2*," press release (London: 12 January 1999); GFMS, "Publication of *Gold Survey 1999–Update 2*," press release (London: 12 January 2000). Data for 1965–99 from IMF, *International Financial Statistics* (Washington, DC: various issues), cited in John E. Young, *Gold: At What Price?* prepared for Mineral Policy Center, Project Underground, and Western Organization of Resource Councils (Berkeley, CA: February 2000, draft).

11. Office of the Treasurer, Commonwealth of Australia, "Reserve Bank of Australia Official Sales of Gold," press release (Canberra: 3 July 1997).

12. Bank of England, "Restructuring the UK's Reserve Holdings: Gold Auctions," press release (London: 7 May 1999); Swiss from Lehman Brothers, op. cit. note 5.

13. Earle Amey, "Gold," in USGS, *Metal Prices in the United States Through 1998* (Reston, VA: 1998).

14. "Central-Bank Gold: Melting Away," *The Economist*, 4 April 1992.

15. U.S. holdings from Lehman Brothers, op. cit. note 5; lost revenue from Dale Henderson and Stephen Salant, "A Note on Government Gold Policies," prepared for Board of Governors of the Federal Reserve System, Washington, DC, 1997.

16. Young, op. cit. note 10; Lehman Brothers, op. cit. note 5.

17. Timbarra from Mineral Policy Center, *Metals Watch*, 28 September 1999; Leadville from Amey, discussion with author, op. cit. note 2.

18. GFMS, "Publication of *Gold Survey 1999–Update 2*," op. cit. note 10.

19. Share of gold from "Exploration Spending Falls," *Financial Times*, 6 January 2000; decline in spending from Metals Economics Group (MEG), "A 23% Decrease in 1999 Exploration Budgets," press release (Halifax, NS, Canada: 20 October 1999); MEG, "A 31% Decrease in 1998 Exploration Budgets," press release (Halifax, NS, Canada: 20 October 1998).

20. George J. Coakley and Thomas B. Dolley, "The Mineral Industry of South Africa," in USGS, *Minerals Information Summary* (Reston, VA: 1996).

21. Amey, op. cit. note 13; Lehman Brothers, op. cit. note 5.

22. Amey, discussion with author, op. cit. note 2; "Gold: Declining Value Sends Shock Waves Through Africa," *UN Wire*, 21 June 1999.

23. Roger Moody, "The Lure of Gold: How Golden is the Future?" Panos Media Briefing No. 9 (London: Panos Institute, May 1996).

24. Waste from John E. Young, "Gold Production at Record High," in Lester R. Brown, Hal Kane, and David Malin Roodman, *Vital Signs 1994* (New York: W.W. Norton & Company, 1994); other damage from Mineral Policy Center, *Golden Dreams, Poisoned Streams* (Washington, DC: 1997).

25. Guy Gugliotta, "'Toxic Bullet' of Cyanide Contaminates Rivers," *Washington Post*, 15 February 2000; Michael J. Jordan, "Epic Poisoning of Europe's Rivers," *Christian Science Monitor*, 16 February 2000; Mineral Policy Center, "Truck Overturns in Route to Kumtor Gold Mine, Spills Sodium Cyanide into Local River," press release (Washington, DC: 4 June 1998).

26. Mineral Policy Center, op. cit. note 24.

TOURISM GROWTH REBOUNDS
(pages 82–83)

1. Rosa Songel, Statistics and Economic Measurement of Tourism, World Tourism Organization (WTO), e-mails to author, 25 November 1999 and 27 January 2000.

2. Ibid.; WTO, "La Región Asia-Pacífico Vuelve a Liderar el Crecimiento del Turismo Mundial," press release (Madrid: 25 January 1999).

3. Songel, op. cit. note 1.

4. Ibid.

5. Ibid.

6. Ibid.

7. Ibid.

8. Ibid.

9. Ibid.; WTO, "Asian Crisis Offers Lessons for Tourism Worldwide," *WTO News*, July/August 1999. Here, Asia includes East Asia and the Pacific but not South Asia

10. Songel, op. cit. note 1.

11. WTO, "Leaner, More Competitive Tourism Sector Emerging: East Asia/Pacific Arrivals Set to Double in Next Ten Years," press release (Madrid: 9 January 1999).

12. WTO, *Tourism Highlights 1999* (Madrid: 19. May 1999); Songel, op. cit. note 1.

13. WTO, op. cit. note 12.

14. WTO, "WTO Picks Hot Tourism Trends for 21st Century," press release (Madrid: 4 June 1998).

15. WTO, op. cit. note 12.

16. Ibid.

17. World Travel and Tourism Council (WTTC), *Travel & Tourism's Economic Impact* (London: March 1999).

18. Ibid.

19. Ibid.

20. Carolyn Cain, consultant for the International Finance Corporation, cited in World Bank, "World Bank Group and World Tourism Organization Examine Role of Tourism in Development," press release (Washington, DC: 25 June 1998).

21. World Bank, op. cit. note 20.

22. Songel, op. cit. note 1.

23. WTTC, op. cit. note 17; Murray Hiebert, "Paradise Lost," *Far Eastern Economic Review*, 23 July 1998.

24. Ramesh Jaura, "Big Increase Forecast in Global Tourism," *Inter Press Service*, 8 March 1999; Birgit Steck, *Sustainable Tourism as a Development Option* (Eschborn, Germany: German Federal Ministry for Economic Co-operation and Development and Deutsche Gesellschaft fur Technische Zusammenarbeit GmbH, 1999).

25. World Bank estimate of 55 percent cited in Alex Markels, "The Great Eco-trips: Guide to the Guides," *Audubon*, September-October 1998.

26. Songel, op. cit. note 1.

27. Francesco Frangialli, "Preserving Paradise," *Our Planet* (U.N. Environment Programme), vol. 10, no. 3 (1999); The Antarctica Project, "Remarkable Lake Vostok: A Case for Strong Protection," *Newsletter*, December 1999.

28. Jonathan Friedland, "Build First, Ask Later: Hotel Boom Threatens Balance in Yucatan," *Wall Street Journal*, 20 August 1999.

29. Jules Siegel, "Mexican President Dives Into Cancun Eco-Protection," *Environment News Service*, 10 September 1999; Michael Christie, "Turtles Face New Danger in Mexico's Caribbean," *Reuters*, 13 November 1999.

30. Steck, op. cit. note 24; WTO, *Guide for Local Authorities on Developing Sustainable Tourism* (Madrid: 1998); James E. N. Sweeting, Aaron G. Bruner, and Amy B. Rosenfeld, "The Green Host Effect: An Integrated Approach to Sustainable Tourism and Resort Development,"

CI Policy Papers (Washington, DC: Conservation International, 1999).

31. Green Hotels Association, "Green Ideas," <www.greenhotels.com/grnideas.htm>, viewed 24 January 2000; International Hotels Environment Initiative, "Six Good Reasons for Going Green," <www.ihei.org>, viewed 24 January 2000; U.N. Environment Programme, "UNEP Hosts Meeting to Develop Tour Operator Initiative for Sustainable Tourism Development," press release (Nairobi: 6 July 1999); WTO, "Tourism Sector Takes Steps to Ensure Future Growth: Global Code of Ethics Adopted at WTO Summit," press release (Madrid: 1 October 1999).

VEHICLE PRODUCTION INCREASES (pages 86–87)

1. Figures for 1995–99 from Standard and Poor's DRI, *World Car Industry Forecast Report, December 1999* (London: 1999); earlier data from American Automobile Manufacturers Association (AAMA), *World Motor Vehicle Facts and Figures 1998* (Washington, DC: 1998).
2. DRI, op. cit. note 1; AAMA, op. cit. note 1.
3. Calculated from DRI, op. cit. note 1. DRI groups Mexico with the United States and Canada in a NAFTA regional group, but here it is included in Latin America.
4. DRI, op. cit. note 1.
5. Ibid.
6. Ibid.
7. Ibid.
8. Colin Couchman, Standard and Poor's DRI, Global Automotive Group, London, e-mail to author, 5 January 2000.
9. People-per-car ratio calculated from DRI fleet data (from *World Car Industry Forecast Report, November 1998*) and world population figure derived from U.S. Bureau of the Census, *International Data Base*, electronic database, Suitland, MD, updated 28 December 1999.
10. DRI, op. cit. note 9; Bureau of Census, op. cit. note 9.
11. Robert M. Heavenrich and Karl H. Hellman, *Light Duty Automotive Technology and Fuel Economy Trends Through 1999* (Washington, DC: Environmental Protection Agency, September 1999).
12. Ibid.
13. Ibid.
14. Ibid.
15. Ibid.
16. Keith Bradsher, "Light Trucks Increase Profits But Foul Air More Than Cars," *New York Times*, 30 November 1997.
17. Matthew L. Wald, "Stricter Pollution Controls Set for Cars and Light Trucks," *New York Times*, 21 December 1999.
18. DRI, op. cit. note 1.
19. Ibid. DRI includes vehicles up to 6 tons in these statistics.
20. John Tagliabue, "Translating S.U.V. Into European," *New York Times*, 14 December 1999.
21. Ibid.
22. Ibid.

BICYCLE PRODUCTION DOWN AGAIN (pages 88–89)

1. "World Market Report," *Interbike Directory 2000* (Laguna Beach, CA: Miller-Freeman, 2000); United Nations, *Industrial Commodity Statistics Yearbook 1997* (New York: United Nations, forthcoming).
2. Worldwatch calculation based on data from "World Market Report," *Interbike Directory* (various issues), and from United Nations, *Industrial Commodity Statistics Yearbook* (New York: United Nations, various issues).
3. "World Market Report," op. cit. note 1.
4. Jay Townley, Bicycle Council, Kent, WA, discussion with author, 27 January 2000.
5. "World Market Report," op. cit. note 1.
6. Ibid.
7. National Sporting Goods Association, <www.nsgs.org/guests/research/participation/partic4.html>, viewed 28 January 2000.
8. "World Market Report," op. cit. note 1.
9. Ibid.
10. Ibid.
11. Frank Jamerson, Electric Battery Bicycle Company, and Ed Benjamin, United States Electric Bicycle and Scooter Association, e-mail to author, 14 January 2000.
12. Ibid.
13. Walter Hook and John Ernst, "Bicycle Use Plunges: The Struggle for Sustainability in China's Cities," *Sustainable Transport*, fall 1999.
14. Ibid.
15. "White Bikes Return to Amsterdam," *Bicycle Retailer and Industry News*, 1 November 1999.
16. Ibid.
17. Sustrans, <www.sustrans.org.uk>, viewed 3 February 2000.
18. Ibid.
19. Ibid.

20. Carlos Alberto Montes, Gerente de Ciclorutas, Instituto de Desarrollo Urbano, Alcaldia Mayor, Santa Fe de Bogota, discussion with author, 14 October 1999.
21. "Bogota Plans Major Bike System Investments," *Sustainable Transport*, fall 1999.
22. The Honorable John Anderson, MP, Minister for Transport and Regional Services, Australia, Speech for the launch of Australia Cycling: The National Strategy 1999-2004, 19 February 1999.

TELEPHONE NETWORK DIVERSIFIES (pages 92–93)

1. International Telecommunication Union (ITU), *World Telecommunication Development Report 1999* (Geneva: 1999).
2. Compound annual growth rate between 1990 and 1998 is 6 percent; ITU, op. cit. note 1; ITU, *World Telecommunication Indicators '98*, Socioeconomic Time-series Access and Retrieval System (STARS) database, downloaded 24 August 1999.
3. ITU, op. cit. note 1.
4. Ibid.; ITU, op. cit. note 2.
5. ITU, op. cit. note 1; ITU, op. cit. note 2.
6. ITU, op. cit. note 1; ITU, *Direction of Traffic, 1999: Trading Telecom Minutes* (Geneva: 1999).
7. Douglas P.E. Smith, *Convergence: Telephones, Mobile Phones and the Internet* (New York: Salomon Brothers, 5 September 1997).
8. Heather E. Hudson, *Global Connections: International Telecommunications Infrastructure and Policy* (New York: Van Nostrand Reinhold, 1997).
9. Annabel Z. Dodd, *The Essential Guide to Telecommunications*, 2nd ed. (Upper Saddle River, NJ: Prentice Hall, 1999); David D. Clark, "High Speed Data Races Home," *Scientific American*, October 1999.
10. Bruce Meyerson, "Cos. Team on Wireless Standards," *Associated Press*, 23 August 1999; Gautam Naik, "That's a WAP: How the Cell Phone and Web Contracted an Arranged Marriage," *Wall Street Journal*, 11 October 1999.
11. ITU, *Trends in Telecommunication Reform: Convergence and Regulation* (Geneva: ITU, 1999).
12. Ibid.
13. Ibid.
14. "Cutting the Cord," Telecommunications Survey, *The Economist*, 9 October 1999.
15. ITU, op. cit. note 1.
16. Ibid.
17. Ibid.
18. Ibid.
19. "The Downside of Mobile," Telecommunications Survey, *The Economist*, 9 October 1999.
20. "Of Mobile Phones and Morbidity," *Environmental Health Perspectives*, October 1998; World Health Organization, "Electromagnetic Fields and Public Health: Mobile Telephones and Their Base Stations," Fact Sheet Number 193, May 1998, <www.who.int/inf-fs/en/fact193. html>.

INTERNET USE ACCELERATES (pages 94–95)

1. Host computer count from Internet Software Consortium and Network Wizards, "Internet Domain Surveys," <www.isc.org/ds>, viewed 20 February 2000. A single host computer can wire several computers to the Internet in the same way that one telephone line can plug in multiple phone extensions. Number of users is a Worldwatch estimate based on Nua Ltd., "How Many Online?" <www.nua.ie>, updated February 2000, and on Computer Industry Almanac, Inc., "U.S. Tops 100 Million Internet Users According to Computer Industry Almanac," press release (Arlington Heights, IL: 10 February 1999). Users are estimated in terms of individuals who get access to the Internet on a weekly basis. Because estimates for the number of users can vary, host computers provide a more reliable measure of the Internet's size.
2. Based on data from Internet Software Consortium and Network Wizards, op. cit. note 1.
3. Computer Industry Almanac, op. cit. note 1.
4. Ibid.; Nua, op. cit. note 1.
5. Computer Industry Almanac, op. cit. note 1.
6. Nua, op. cit. note 1.
7. Computer Industry Almanac, op. cit. note 1; Computer Industry Almanac, *Internet User Forecast, 1990–2005* (Arlington Heights, IL: 1999).
8. Computer Industry Almanac, op. cit. note 1.
9. Internet Software Consortium and Network Wizards, op. cit. note 1.
10. Nua, op. cit. note 1.
11. Ibid.; Internet Software Consortium and Network Wizards, op. cit. note 1.
12. Internet Software Consortium and Network Wizards, op. cit. note 1.
13. Ibid.
14. Ibid.
15. Ibid.
16. Ibid.; 1999 was the first year these countries registered their own domain names; prior to

that, they may have had some Internet access through servers in neighboring countries.

17. Internet Software Consortium and Network Wizards, op. cit. note 1; International Telecommunication Union (ITU), *World Telecommunications Indicators on Diskette* (Geneva: 1996); U.S. Bureau of the Census, *International Data Base*, electronic database, Suitland, MD, updated 28 December 1999. In Figure 2, U.S. share of <.com>, <.net>, and <.org> addresses is estimated using data provided by Cheryl Regan, Network Solutions, Herndon, VA, e-mail to author, 29 February 2000.

18. Internet Software Consortium and Network Wizards, op. cit. note 1.

19. Ibid.; Nua, op. cit. note 1.

20. Internet Software Consortium and Network Wizards, op. cit. note 1; Nua, op. cit. note 1.

21. Internet Software Consortium and Network Wizards, op. cit. note 1; Nua, op. cit. note 1; Computer Industry Almanac, op. cit. note 1.

22. Global Reach, "Global Internet Statistics (by Language)," <www/euromarketing.com/globa stats>, updated 20 February 2000.

23. David Lake, "The Web: Growing by 2 Million Pages a Day," *The Industry Standard*, 28 February 2000.

24. Ibid.

25. Stacy Lawrence, "The Net World in Numbers," *The Industry Standard*, 7 February 2000.

26. Travel Industry Association of America, "Internet Usage by Travelers Continues to Soar," news release (Washington, DC: 8 February 2000).

27. Forrester Research, "Online Advertising to Reach $33 Billion Worldwide by 2004," press release (Cambridge, MA: 12 August 2000).

28. Nua Ltd., "Huge October Ad Spend for Dot-coms," 11 February 2000, <www.nua.ie/surveys>.

29. Joseph Romm, Arthur Rosenfeld, and Susan Hermann, *The Internet Economy and Global Warming* (Washington, DC: Center for Energy and Climate Solutions, December 1999).

30. Molly O'Meara, "Harnessing Information Technologies for the Environment," in Lester R. Brown et al., *State of the World 2000* (New York: W.W. Norton & Company, 2000).

WORLD POPULATION PASSES 6 BILLION (pages 98–99)

1. October 12 from United Nations, Population Division, "The World at Six Billion," press release, <www.popin.org/6billion>, 12 October 1999; projections by other organizations, such as the U.S. Bureau of the Census, suggest slightly different dates; 1960 population from United Nations, *World Population Prospects: The 1998 Revision* (New York: December 1998).

2. U.S. Bureau of the Census, *International Data Base*, electronic database, Suitland, MD, updated 28 December 1999.

3. United Nations, *The 1998 Revision*, op. cit. note 1; Barbara Crossette, "In Days, India, Chasing China, Will Have a Billion People," *New York Times*, 5 August 1999.

4. United Nations, *The 1998 Revision*, op. cit. note 1.

5. Bureau of Census, op. cit. note 2.

6. Ibid.

7. Estimate based on United Nations data from Stan Bernstein, Senior Research Adviser, U.N. Population Fund (UNFPA), e-mail to author, 14 January 2000; young men and women entering reproductive age is defined as those now between the ages of 15 and 24.

8. United Nations, *The 1998 Revision*, op. cit. note 1.

9. Population Reference Bureau, "1999 World Population Datasheet," wall chart (Washington, DC: June 1999).

10. United Nations, *The 1998 Revision*, op. cit. note 1.

11. UNFPA, *The State of World Population 1999* (New York: 1999).

12. Ibid.

13. Ibid.

14. Nancy E. Riley, "Gender, Power, and Population Change," *Population Bulletin*, May 1997.

15. Sandra Postel, *Pillar of Sand* (New York: W.W. Norton & Company, 1999); water scarcity is defined as less than 1,700 cubic meters of annual freshwater runoff per person, a quantity that most water analysts feel is necessary to meet basic agricultural, domestic, and hygienic needs.

16. Worldwatch estimate based on United Nations, *The 1998 Revision*, op. cit. note 1.

17. United Nations, "Key Actions for the Further Implementation of the Programme of Action of the International Conference on Population and Development," Twenty-first Special Session of the United Nations General Assembly (New York: 1 July 1999).

18. UNFPA, op. cit. note 11.

HIV/AIDS PANDEMIC HITS AFRICA HARDEST (pages 100–01)

1. Joint United Nations Programme on HIV/AIDS (UNAIDS), *AIDS Epidemic Update: December 1999* (Geneva: 1999); number of new infections updated by Neff Walker, UNAIDS, Geneva, e-mail to author, 20 March 2000.
2. UNAIDS, op. cit. note 1.
3. Ibid.; New York City from United Nations, *World Urbanization Prospects: The 1996 Revision* (New York: 1998).
4. Barbara Crossette, "Gore Presides Over Rare Security Council Debate on AIDS," *New York Times*, 11 January 2000.
5. UNAIDS, op. cit. note 1.
6. Ibid.
7. Myron Essex, "The New AIDS Epidemic," *Harvard Magazine*, September-October 1999; UNAIDS, *Report on the Global HIV/AIDS Epidemic: June 1998* (Geneva: 1998); "35% of Namibians Have HIV-AIDS," *The Namibian*, 6 December 1999.
8. UNAIDS, op. cit. note 7.
9. UNAIDS, op. cit. note 1.
10. Ibid.
11. Ibid.
12. Ibid.; UNAIDS, op. cit. note 7.
13. UNAIDS, op. cit. note 1.
14. Prevalence and number infected in 1999 from ibid.; susceptible population from Population Reference Bureau, "1999 World Population Datasheet," wall chart (Washington, DC: June 1999).
15. UNAIDS, op. cit. note 1.
16. Ibid.
17. UNICEF, *Children Orphaned by AIDS: Front-line Responses from Eastern and Southern Africa* (New York: December 1999).
18. Ibid.
19. Princeton N. Lyman, "Facing a Global AIDS Crisis," *Washington Post*, 11 August 1999.
20. World Bank, *Confronting AIDS: Public Priorities in a Global Epidemic* (New York: Oxford University Press, 1997).
21. Robert S.-H. Shell, "Halfway to the Holocaust," Population Research Unit, Rhodes University, East London, South Africa, unpublished manuscript.
22. UNICEF, op. cit. note 17.
23. Mercedes Sayagues, "How AIDS is Starving Zimbabwe," *Daily Mail & Guardian* (Johannesburg), 16 August 1999.
24. "Level and Flow of National and International Resources for the Response to HIV/AIDS, 1996–1997," press release, UNAIDS, Geneva, 22 April 1999.

REFUGEE NUMBERS CONTINUE DECLINE (pages 102–03)

1. U.N. High Commissioner for Refugees (UNHCR), "UNHCR by Numbers," Table 1, <www.unhcr.ch/un&ref/numbers/table1.htm>, viewed 29 December 1999.
2. Ibid.; comparison with 1995 from Jennifer D. Mitchell, "Refugee Flows Drop Steeply," in Lester R. Brown, Michael Renner, and Christopher Flavin, *Vital Signs 1998* (New York: W.W. Norton & Company, 1998).
3. UNHCR, "Refugees and Others of Concern to UNHCR. 1998 Statistical Overview," <www.unhcr.ch/statist/98oview/>, viewed 29 December 1999.
4. U.S. Committee for Refugees, *World Refugee Survey 1999* (Washington, DC: 1999).
5. UNHCR, op. cit. note 3.
6. Ibid.
7. Ibid.
8. Ibid.
9. UNHCR, op. cit. note 1.
10. Ibid.
11. UNHCR's mandate defines refugees as those "who are outside their countries and who cannot or do not want to return because of a well-founded fear of being persecuted for reasons of their race, religion, nationality, political opinion or membership in a particular social group." UNHCR, "UNHCR by Numbers," <www.unhcr.ch/un&ref/>, viewed 29 December 1999.
12. U.S. Committee for Refugees, op. cit. note 4.
13. Ibid.
14. Ibid.
15. Ibid.
16. UNHCR, op. cit. note 1.
17. U.S. Committee for Refugees, op. cit. note 4.
18. The figure of 57 million is arrived at by adding the populations of concern to UNHCR (excluding returnees, 19.1 million), Palestinians cared for by UNRWA (3.2 million), the internally displaced (some 30 million), and those in refugee-like situations (5 million).
19. UNHCR, op. cit. note 3.
20. U.S. Committee for Refugees, op. cit. note 4.
21. Ibid.
22. Ibid.

URBAN POPULATION CONTINUES TO RISE (pages 104–05)

1. Preliminary 1999 estimate from United Nations, *World Urbanization Prospects: The 1999 Revision* (New York: in press), and from U.N. Population Division, New York, e-mail to author, 1 December 1999; 1996 estimate from United Nations, *World Population Prospects: The 1996 Revision* (New York: 1998).
2. United Nations, *The 1996 Revision*, op. cit. note 1.
3. U.N. Population Division, op. cit. note 1.
4. United Nations, *The 1996 Revision*, op. cit. note 1.
5. Molly O'Meara, *Reinventing Cities for People and the Planet*, Worldwatch Paper 147 (Washington, DC: Worldwatch Institute, June 1999).
6. Calculation based on urban population from United Nations, *The 1996 Revision*, op. cit. note 1; share of GDP from industry and services from World Bank, *World Development Indicators 1997*, on CD-ROM (Washington, DC: 1997), and from World Bank, *World Bank Indicators 1998* (Washington, DC: 1998); carbon emissions from G. Marland et al., "Global, Regional, and National CO_2 Emission Estimates from Fossil Fuel Burning, Cement Production, and Gas Flaring: 1751–1995 (revised March 1999)," Oak Ridge National Laboratory, <cdiac.esd.ornl.gov>, viewed 22 April 1999; industrial roundwood consumption from U.N. Food and Agriculture Organization, *FAOSTAT Statistics Database*, <apps.fao.org>.
7. United Nations, *The 1996 Revision*, op. cit. note 1.
8. Mathias Wackernagel and William Rees, *Our Ecological Footprint: Reducing Human Impact on the Earth* (Philadelphia, PA: New Society Publishers, 1996).
9. Calculation based on Herbert Girardet, "Cities and the Biosphere," UNDP Roundtable, The Next Millennium: Cities for People in a Globalizing World, Marmaris, Turkey, 19–21 April 1996; International Institute for Environment and Development, for the U.K. Department of Environment, *Citizen Action to Lighten Britain's Ecological Footprints* (London: 1995).
10. Tertius Chandler, *Four Thousand Years of Urban Growth: An Historical Census* (Lewiston, NY: Edwin Mellen Press, 1987).
11. United Nations, *The 1996 Revision*, op. cit. note 1.
12. Ibid.
13. Ibid.
14. U.N. Centre for Human Settlements (Habitat), *An Urbanizing World: Global Report on Human Settlements 1996* (Oxford, U.K.: Oxford University Press, 1996).
15. World Resources Institute et al., *World Resources 1996–97* (New York: Oxford University Press, 1996).
16. Anne Whiston Spirn, *The Granite Garden: Urban Nature and Human Design* (New York: Basic Books, 1984); Herbert Girardet, *Creating Sustainable Cities*, Schumacher Briefing No. 2 (Devon, U.K.: Green Books, 1999).

CIGARETTE DEATH TOLL RISING (pages 106–07)

1. World Health Organization (WHO), *The World Health Report 1999: Making a Difference* (Geneva: 1999).
2. U.S. Department of Agriculture (USDA), Foreign Agriculture Service (FAS), *Special Report: World Cigarette Situation* (Washington, DC: August 1999).
3. Population from U.S. Bureau of the Census, *International Data Base*, electronic database, Suitland, MD, updated 28 December 1999.
4. "Tobacco: WHO Chief Urges a Global Campaign Against Smoking," *UN Wire*, 19 March 1999.
5. WHO, op. cit. note 1.
6. Ibid.
7. Ibid.
8. Figure of $200 billion from "Worldwide Tobacco Use on the Rise," *China Daily*, 11 November 1999; $20 billion from "Tobacco: Industry Hides the Ill-Effects of Smoking, WHO Charges," *UN Wire*, 11 May 1999.
9. USDA, op. cit. note 2.
10. Corinne Podger, "Smoking Set To Kill Millions in China," *BBC News*, 5 October 1999.
11. Ibid.
12. Ibid.
13. USDA, op. cit. note 2.
14. Ibid.
15. Cigarette consumption data from Arnella Trent, Tobacco Analyst, USDA, FAS, Washington, DC, letter to author, 22 December 1999.
16. Peter Ford, "Europe Takes on Tobacco," *Christian Science Monitor*, 22 September 1999.
17. Ibid.
18. U.S. consumption from Trent, op. cit. note 15, and population from Bureau of Census, op. cit. note 3.
19. WHO, op. cit. note 1.
20. Scott Gold, "Venezuela Sues U.S. Tobacco Firms," *The Oregonian*, 28 January 1999.
21. Ibid.

22. USDA, op. cit. note 2.
23. Saundra Torry, "Foreign Nations Sue Tobacco Companies," *The Oregonian*, 19 January 1999; Ford, op. cit. note 16.
24. USDA, Economic Research Service, "Cigarette Price Increase Follows Tobacco Pact," *Agricultural Outlook*, January-February 1999.
25. WHO, "The Framework Convention on Tobacco Control," <www.who.int/toh/fctc/fctcintro.htm>, viewed 21 December 1999.
26. "Anti-Smoking Goes Global," *Washington Post*, 10 July 1999.

NUMBER OF WARS ON UPSWING
(pages 110–11)

1. Arbeitsgemeinschaft Kriegsursachenforschung (AKUF), "Weltweit 35 Kriege in 1999. Erneuter Anstieg der Kriegszahl," press release (Hamburg, Germany: Institute for Political Science, University of Hamburg, 20 December 1999). Due to definitional and methodological reasons, other analysts—at the University of Uppsala, Sweden, and Project Ploughshares, Canada—report somewhat higher numbers.
2. AKUF, op. cit. note 1.
3. Worldwatch calculation, based on ibid., on AKUF, "Kriegesbilanz 1998," press release (Hamburg, Germany: Institute for Political Science, University of Hamburg, revised 11 January 1999), and on AKUF, "Kriege und Bewaffnete Konflikte 1997," <www.sozialwiss.uni-hamburg.de/Ipw/Akuf/kriege_text.html>, viewed 4 January 1999.
4. Calculated from AKUF, op. cit. note 1.
5. Michael Renner, *Ending Violent Conflict*, Worldwatch Paper 146 (Washington, DC: Worldwatch Institute, April 1999).
6. AKUF, op. cit. note 1.
7. Ibid.
8. Margareta Sollenberg, ed., *States in Armed Conflict 1998*, Report No. 54 (Uppsala, Sweden: Uppsala University, Department of Peace and Conflict Research, 1999).
9. Ibid.
10. Ibid.
11. Ibid.
12. AKUF, op. cit. note 1.
13. Ibid.
14. Sollenberg, op. cit. note 8.
15. Project Ploughshares, *Armed Conflicts Report '99* (Waterloo, Canada: Institute of Peace and Conflict Studies, summer 1999).
16. Ibid.
17. Ibid.
18. Ibid.
19. Ibid.

PEACEKEEPING EXPENDITURES TURN UP (pages 112–13)

1. Bock Cheng Yeo, Officer in Charge, Peacekeeping Finance Division, United Nations, e mail to author, 19 January 2000; U.N. Department of Public Information (UNDPI), "United Nations Peacekeeping Operations. Background Note," New York, 1 February 2000; Wasantha Bandarage, UNDPI, discussion with author, 23 December 1999. (Beginning with July 1997, the United Nations switched its peacekeeping accounts from calendar years to July–June reporting periods.) The increase in appropriations is due to the likely expansion of missions in East Timor, Sierra Leone, and the Democratic Republic of Congo.
2. UNDPI, op. cit. note 1.
3. Ibid.
4. U.N. Department of Peacekeeping Operations, "Current Peacekeeping Operations," <www.un.org/Depts/DPKO/c_miss.htm>, viewed 17 December 1999.
5. U.N. Department of Peacekeeping Operations, "Completed Peacekeeping Operations," <www.un.org/Depts/DPKO/p_miss.htm>, viewed 17 December 1999.
6. U.N. Department of Peacekeeping Operations, op. cit. note 4; "Security Council Expands Sierra Leone Force to 11,100," *United Nations Newservice*, 7 February 2000.
7. U.N. Department of Peacekeeping Operations, op. cit. note 4. The current East Timor operation, UNTAET (U.N. Transitional Administration in East Timor), is a follow-on to the U.N. Mission in East Timor (UNAMET) that was established to assist in the process by which the population of East Timor was to decide whether to accept autonomy status within Indonesia or to pursue independence. UNAMET lasted from 11 June 1999 to 25 October 1999. See U.N. Department of Peacekeeping Operations, "UNTAET," <www.un.org/peace/etimor/UntaetB.htm>, viewed 17 December 1999.
8. UNDPI, op. cit. note 1.
9. Calculated from U.N. Department of Peacekeeping Operations, op. cit. note 5, and from U.N. Department of Peacekeeping Operations, op. cit. note 4. Some conflicts have

attracted a succession of U.N. missions. Eight missions were dispatched to different parts of the former Yugoslavia, and four each to Angola and Haiti.

10. "Status of Contributions to the Regular Budget, International Tribunals and Peacekeeping Operations as at 15 December 1999," Office of the Spokesman for the Secretary-General, United Nations, New York.

11. Ibid.

12. The NATO air war against Serbia that led to the establishment of KFOR cost at least another $11 billion; International Institute for Strategic Studies, *The Military Balance 1999/2000* (London: Oxford University Press, 1999).

13. Worldwatch Institute estimate based on various sources, including Johan Sjöberg, "Multilateral Peace Missions, 1998," in Stockholm International Peace Research Institute, *SIPRI Yearbook 1999. Armaments, Disarmament and International Security* (New York: Oxford University Press, 1999), and International Institute for Strategic Studies, op. cit. note 12.

14. Ibid.

15. Ibid.

TRANSGENIC CROP AREA SURGES (pages 118–19)

1. Clive James, *Global Review of Commercialized Transgenic Crops: 1999 (Preview)*, ISAAA Brief No. 12 (Ithaca, NY: International Service for the Acquisition of Agri-Biotech Applications, 1999).

2. Ibid.

3. Ibid.

4. Ibid.

5. Michael Byrnes, "World GM Crop Sales Up to US$3 BLN in '99" *Reuters*, 2 March 2000.

6. James, op. cit. note 1.

8. Ibid.

8. Ibid.

9. Ibid.

10. Janet Carpenter and Leonard Gianessi, "Herbicide Use on Roundup Ready Crops," *Science*, 4 February 2000; Janet E. Carpenter and Leonard P. Gianessi, *Value of Bt and Herbicide-Resistant Cottons* (Washington, DC: National Center for Food and Agricultural Policy, forthcoming); Charles Benbrook, "World Food System Challenges and Opportunities: GMOs, Biodiversity, and Lessons from America's Heartland," comments before the University of Illinois World Food and Sustainable Agriculture Program, 27 January 1999, < www.biotech-

info.net/IWFS.pdf >, viewed 15 February 1999; Jeffrey Hyde et al., "The Economics of Bt-Corn: Adoption Implications," Purdue University Cooperative Extension Service, West Lafayette, IN, August 1998.

11. S.W.B. Ewan and A. Pusztai, "Effects of Diets Containing Genetically Modified Potatoes Expressing Galanthus Nivalis Lectin on Rat Small Intensine," *The Lancet*, 16 October 1999.

12. J.J.E. Losey, L.S. Rayor, and M.E. Carter, "Transgenic Pollen Harms Monarch Larvae," *Nature*, 20 May 1999; D. Saxena, S. Flores, and G. Stotzky, "Insecticidal Toxin in Root Exudates from Bt Corn," *Nature*, 2 December 1999.

13. Barnaby J. Feder, "Monsanto to Bar a Class of Seeds," *New York Times*, 5 October 1999; Rural Advancement Foundation International, "Suicide Seeds on the Fast Track," news release (Carrboro, NC: 25 February 2000).

14. Scott Kilman, "Monsanto is Sued Over Genetically Altered Crops," *Wall Street Journal*, 15 December 1999.

15. GeneWatch (UK), *GM Crops and Foods: A Review of Developments in 1999*, Briefing No. 9 (Derbyshire: January 2000).

16. Ibid.

17. Kirsten Dawkins, *Biotech—Seattle to Montreal and Beyond: The Battle Royale of the 21st Century* (Minneapolis, MN: Institute for Agriculture and Trade Policy, forthcoming).

18. "To Plant or Not to Plant," *The Economist*, 15 January 2000; U.S. Department of Agriculture (USDA), Foreign Agricultural Service, "U.S. Export Sales Reports," < www.fas.usda.gov/export-sales >, viewed 15 January 2000.

19. "New Genes Meet a Wary Market," *Christian Science Monitor*, 8 December 1999.

20. David J. Morrow and Andrew Ross Sorkin, "2 Drug Companies to Combine Troubled Agricultural Units," *New York Times*, 2 December 1999; David Barboza, "Monsanto and Pharmacia to Join, Creating a Pharmaceutical Giant," *New York Times*, 20 December 1999.

21. USDA, Economic Research Service, "Impacts of Adopting Genetically Engineered Crops in the U.S.—Preliminary Results," 20 July 1999, < www.econ.ag.gov/whatsnew/issues/gmo >, viewed 20 December 1999; Carpenter and Gianessi, "Herbicide Use on Roundup Ready Crops," op. cit. note 10; Carpenter and Gianessi, *Value of Bt and Herbicide-Resistant Cottons*, op. cit. note 10; Charles Benbrook, "Evidence of the Magnitude of the Roundup Ready Soybean

Yield Drag from University-Based Varietal Trials in 1998," Ag BioTech InfoNet Technical Paper No. 1, 13 July 1999, <www.biotech-info.net/RR_yield_drag_98.pdf>, viewed 20 December 1999.

22. USDA, op. cit. note 21; Carpenter and Gianessi, *Value of Bt and Herbicide-Resistant Cottons,* op. cit. note 10.

23. Bryan J. Hubbell and Rick Welsh, "Transgenic Crops. Engineering a More Sustainable Agriculture," *Agriculture and Human Values,* October 1998; Charles Benbrook, "Who Controls and Who Will Benefit from Plant Genomics?" presented at The 2000 Genome Seminar: Genomic Revolution in the Fields: Facing the Needs of the New Millennium, American Association for the Advancement of Science Annual Meeting, Washington, DC, 17–22 February 2000.

24. Guy Gugliotta, "Gene-Altered Rice May Help Fight Vitamin A Deficiency Globally," *Washington Post,* 14 January 2000.

25. Vandana Shiva, "Genetically Engineered Vitamin A Rice: A Blind Approach to Blindness Prevention," e-mail to author, 14 February 2000; Miguel Altieri and Peter Rosset, "Ten Reasons Why Biotechnology Will Not Ensure Food Security, Protect the Environment and Reduce Poverty in the Developing World" (Oakland, CA: Institute for Food and Development Policy/Food First, October 1999).

26. Andrew Pollack, "130 Nations Agree on Safety Rules for Biotech Foods," *New York Times,* 30 January 2000; "Cartagena Protocol on Biosafety to the Convention on Biological Diversity," final text of 23 February 2000, at <www.biodiv.org/ biosafe/BIOSAFETY-PROTOCOL.htm>. In addition to transgenic crops, the protocol covers all other "living modified organisms," including transgenic fish, trees, and bacteria.

27. Pollack, op. cit. note 26; "Cartagena Protocol," op. cit. note 26.

28. Labeling bills from Marian Burros, "Documents Show Officials Disagreed on Altered Food," *New York Times,* 1 December 1999, and from "Boxer Calls for Labeling Genetically Engineered Food," press release (Washington, DC: Office of Senator Barbara Boxer, 22 February 2000); "US to Look at Biotech Foods as World Concern Increases," *Reuters,* 19 October 1999.

29. American Corn Growers Association, "Corn Growers Complete Survey on Farmer Planting Intentions for Upcoming Growing Season," press release (Washington, DC: 21 February 2000); Randy Fabi, "U.S. Farmers Plan Decline in Biotech Crops," *Reuters,* 13 January 2000.

ORGANIC FARMING THRIVES WORLDWIDE (pages 120–21)

1. International Trade Centre (ITC), *Organic Food and Beverages: World Supply and Major European Market* (Geneva: U.N. Conference on Trade and Development, 1999).

2. Bernward Geier, Executive Director, International Federation of Organic Agriculture Movements (IFOAM), e-mail to author, 23 January 2000.

3. Area is Worldwatch estimate based on Helga Willer and Minou Yussefi, *Organic Agriculture World-Wide* (Bad Dürkheim: Stiftung Ökologie & Landbau, 2000), and on Catherine Greene, Economic Research Service (ERS), U.S. Department of Agriculture (USDA), discussion with author, 11 January 2000; sales is Worldwatch estimate based on Willer and Yussefi, op. cit. this note, and on ITC, op. cit. note 1.

4. The term "certified organic" indicates that some private or public certifying body has inspected the farm system for adherence to organic practices and principles. Data on certified organic area is used in this analysis. While a diversity of farming systems, especially in the developing world, might qualify as organic, lack of certification means that they are not officially counted as organic.

5. Data provided by Nic Lampkin, University of Wales, Aberystwyth, e-mail to author, 25 January 2000. This includes area "in conversion" or "transition": when a farm that has been using conventional, chemical methods begins to practice organic methods, it cannot advertise its products as certified organic; this period ranges from three to five years.

6. Nicolas Lampkin, "Organic Farming in the European Union—Overview, Policies and Perspectives," paper presented at Conference on Organic Farming in the European Union—Perspectives for the 21st Century, Vienna, 27–28 May 1999.

7. Lampkin, op. cit. note 5.

8. Geier, op. cit. note 2.

9. Canada from Willer and Yussefi, op. cit. note 3; Australia from Liz Clay, "Organic Trade in Oceania," in Willie Lockeretz and Bernward Geier, eds., *Quality and Communication for the Organic Market,* proceedings of the Sixth IFOAM Trade Conference (Tholey-Theley, Germany:

IFOAM, 2000); Japan from Fumiko Masuda, "The Domestic Organic Market and the Development of National Standards in Asia," in ibid.

10. Growth rate and area estimate for the United States is for 1997 and is from Catherine Greene, ERS, USDA, discussion with author, 27 January 2000; Canada from Willer and Yussefi, op. cit. note 3.

11. Sales growth from Holly Gibbons, Organic Trade Association, discussion with author, 28 January 2000; size of North America market from Willer and Yussefi, op. cit. note 3.

12. Lampkin, op. cit. note 6.

13. Mark Lipson, Searching for the "O-Word": An Analysis of the USDA Current Research Information System (CRIS) for Pertinence to Organic Farming (Santa Cruz, CA: Organic Farming Research Foundation, 1997).

14. Robert Pear, "Tougher Labeling for Organic Food," New York Times, 11 May 1998.

15. Lampkin, op. cit. note 6.

16. Sarah Ryle, "'Frankenstein' Drives Demand for Organics," The Observer (U.K.), 21 February 1999; Jackie Storer, "Brown Announces 10M Extra Aid for Organic Farming," PA News (U.K.), 28 October 1999.

17. Lampkin, op. cit. note 5.

18. ITC, op. cit. note 1; Willer and Yussefi, op. cit. note 3. Since many farmers in Africa, Asia, and Latin America cannot afford agrochemical inputs, much of the land under cultivation there would qualify as organic. (Such traditional production would likely benefit from the more proactive aspects of organic farming.) These farms are generally not certified because awareness of organic farming is low or certification is too expensive.

19. Willer and Yussefi, op. cit. note 3.

20. Gary C. Groves, "Update on Argentina's Organic Sector" (Washington, DC: Foreign Agricultural Service, USDA, 15 October 1998).

21. Charles Walaga, "Organic Agriculture Trade: State of the Art in Africa," in Lockeretz and Geier, op. cit. note 9.

22. Ibid.

23. Geier, op. cit. note 2.

24. Catherine Murphy, Cultivating Havana: Urban Agriculture and Food Security in the Years of Crisis, Development Report (Oakland, CA: Institute for Food and Development Policy (IFDP)/Food First, May 1999).

25. Proceedings from USDA, ERS, Farm Foundation Workshop, "The Economics of Organic Farming Systems: What Can Long-Term Cropping Systems Studies Tell US?" Washington, DC, 21 April 1999 (forthcoming); Rick Welsh, The Economics of Organic Grain and Soybean Production in the Midwestern United States, Policy Studies Report No. 13 (Greenbelt, MD: Henry A. Wallace Institute for Alternative Agriculture, May 1999); Sean Clark et al, "Crop-yield and Economic Comparisons of Organic, Low-input, and Conventional Farming Systems in California's Sacramento Valley," and Yetunde O. Ogini et al., "Comparison of Organic and Conventional Dairy Farms in Ontario," American Journal of Alternative Agriculture, fall 1999.

26. Welsh, op. cit. note 25; Miguel Altieri, Peter Rosset, and Lori Ann Thrupp, "The Potential for Agroecology to Combat Hunger in the Developing World," Policy Brief (Oakland, CA: IFDP/Food First, October 1998).

27. D. Tilman, "The Greening of the Green Revolution," Nature, 19 November 1998; L.E. Drinkwater, P. Wagoner, and M. Sarrantonio, "Legume-Based Cropping Systems Have Reduced Carbon and Nitrogen Losses," Nature, 19 November 1998; Dick Cobb et al., "Organic Farming Study," Global Environmental Change Programme Briefing No. 17, University of Sussex, March 1998; Soil Association, Biodiversity Benefits of Organic Farming (Bristol, U.K.: April 2000).

28. "Organic Agriculture is Essential for Conserving Biodiversity and Nature," press release for the Vignola Declaration and Action Plan (Tholey-Theley, Germany: IFOAM, 23 May 1999).

29. Farm worker risks from Welsh, op. cit. note 25, and from World Health Organization, The World Health Report 1998 (Geneva: 1998); consumer risks and booming babyfood market from "Greener Greens?" Consumer Reports, January 1998, from J. Michael Harris, "Consumers Pay a Premium for Organic Baby Foods," Food Review, May–August 1997, and from Laura Scandurra, "An Overview of the European Organic Food Market," in Lockeretz and Geier, op. cit. note 9.

30. Willer and Yussefi, op. cit. note 3.

31. U.N. Food and Agriculture Organization, Committee on Agriculture, "Organic Agriculture," position paper, Rome, 25–29 January 1999, viewed at < www.fao.org/WAICENT/FAOINFO/SUSTDEV/Epdirect/Epre0072.htm >, viewed 29 February 2000.

32. Lampkin, op. cit. note 6.

33. Marian Burros, "Strict Rules to Limit Genetic Engineering on Organic Foods," New York Times,

5 March 2000; Catherine Greene, ERS, USDA, "Organic Farming in the US: Farm and Farmer Patterns," presentation at "Organic Agriculture: Growth of Global Markets and Directions for Research," American Association for the Advancement of Science Annual Meeting, Washington, DC, 17–22 February 2000.

34. Andrea Adelson, "Organic Clothing: On Backs, Not Minds," *New York Times*, 6 November 1997; John Fetto, "Home on the Organic Range," *American Demographics*, August 1999.

GROUNDWATER DEPLETION WIDESPREAD (pages 122–23)

1. Sandra Postel, *Pillar of Sand* (New York: W.W. Norton & Company, 1999).
2. Indian Water Resources Society as cited in M. Ramon Llamas, "Considerations on Ethical Issues in Relation to Groundwater Development and/or Mining," paper prepared for the International Conference on Regional Aquifer Systems in Arid Zones: Managing Non-Renewable Resources, Tripoli, Libya, 20–24 November 1999.
3. Ibid.
4. Ministry of Water Resources and Electric Power, People's Republic of China, *Irrigation and Drainage in China* (Beijing: China Water Resources and Electric Power Press, 1987).
5. Ruth Meinzen-Dick, *Groundwater Markets in Pakistan: Participation and Productivity*, Research Report 105 (Washington, DC: International Food Policy Research Institute, 1996).
6. National Research Council, *A New Era for Irrigation* (Washington, DC: National Academy Press, 1996).
7. Ibid.
8. National Environmental Engineering Research Institute, "Water Resources Management in India: Present Status and Solution Paradigm" (Nagpur, India: undated (circa 1997)).
9. Ibid.
10. Zhang Qishun and Zhang Xiao, "Water Issues and Sustainable Social Development in China," *Water International*, vol. 20, no. 3 (1995).
11. Postel, op. cit. note 1.
12. Projected deficit from Sandia National Laboratories, China Infrastructure Initiative: Decision Support Systems, < www.igaia.sandia.gov/igaia/china/chinamodel.html >.
13. Depletion estimates based on data in Edwin D. Gutentag et al., *Geohydrology of the High Plains Aquifer in Parts of Colorado, Kansas, Nebraska, New Mexico, Oklahoma, South Dakota, Texas, and Wyoming* (Washington, DC: U.S. Government Printing Office, 1984), and in Dork L. Sahagian, Frank W. Schwartz, and David K. Jacobs, "Direct Anthropogenic Contributions to Sea Level Rise in the Twentieth Century," *Nature*, 6 January 1994.
14. Depletion estimates abased on Gutentag et al., op. cit. note 13, and on Sahagian, Schwartz, and Jacobs, op. cit. note 13.
15. Depletion estimates abased on Gutentag et al., op. cit. note 13, and on Sahagian, Schwartz, and Jacobs, op. cit. note 13.
16. Llamas, op. cit. note 2.
17. Postel, op. cit. note 1.
18. Llamas, op. cit. note 2.
19. Sandra Postel, *Last Oasis*, rev. ed. (New York: W.W. Norton & Company, 1997).

GROUNDWATER QUALITY DETERIORIATING (pages 124–25)

1. Figure of 97 percent from U.N. Environment Programme (UNEP), *Groundwater: A Threatened Resource* (Nairobi: 1996); drinking water figure based on data on the United States from U.S. Environmental Protection Agency (EPA), Office of Water, *The Quality of Our Nation's Water* (Washington, DC: 1998), on Australia from Environment Australia, *State of the Environment Report 1996* (Canberra: 1996), on Asia-Pacific and Latin America from UNEP, op. cit. this note, and on Europe from Organisation for Economic Co-operation and Development, *Water Resources Management: Integrated Policies* (Paris: 1989).
2. UNEP, op. cit. note 1.
3. International Water Management Institute (IWMI), unpublished data, supplied by Charlotte de Fraiture, IWMI, e-mail to author, 21 October 1999.
4. Francis H. Chapelle, *The Hidden Sea: Ground Water, Springs, and Wells* (Tucson, AZ: Geoscience Press, Inc., 1997).
5. Table 1 based on multiple sources, including U.S. Geological Survey (USGS), *The Quality of Our Nation's Waters—Nutrients and Pesticides* (Reston, VA: 1999); British Geological Survey (BGS) et al., *Characterisation and Assessment of Groundwater Quality Concerns in Asia-Pacific Region* (Oxfordshire, UK: 1996); European Environmental Agency, *Groundwater Quality and Quantity in Europe* (Copenhagen: 1999); and UNEP, op. cit. note 1.

6. BGS et al., op. cit. note 5; World Health Organization (WHO) guidelines reprinted in UNEP, op. cit. note 1.

7. W.L. Zhang et al., "Nitrate Pollution of Groundwater in Northern China," *Agriculture, Ecosystems and Environment*, vol. 59 (1996).

8. Ibid.

9. Linda Nash, "Water Quality and Health," in Peter H. Gleick, ed., *Water in Crisis* (New York: Oxford University Press, 1993).

10. EPA, op. cit. note 1.

11. Ibid.

12. Ross Nickson et al., "Arsenic Poisoning of Bangladesh Groundwater," *Nature*, 24 September 1998; Badal K. Mandal et al., "Arsenic in Groundwater in Seven Districts of West Bengal, India—The Biggest Arsenic Calamity in the World," *Current Science*, 10 June 1996; Stephen Foster, BGS, Oxfordshire, U.K., discussion with author, 3 December 1999.

13. S.K. Acharyya et al., and Tarit Roy Chowdhury et al., "Arsenic Poisoning in the Ganges Delta," *Nature*, 7 October 1999; WHO, "Arsenic in Drinking Water," Fact Sheet No. 210 (Geneva: February 1999).

14. BGS et al., op. cit. note 5.

15. Ibid.

16. WHO from ibid.

17. Bruce Kabelski, EPA, Washington, DC, discussion with author, November 18, 1999; Donald Sutherland, "60 Percent of America's Liquid Toxic Waste Injected Underground," *Environmental News Service*, 7 July 1999.

18. Sutherland, op. cit. note 18.

19. UNEP, op. cit. note 1.

20. Residence time for groundwater from UNEP, op. cit. note 1; time for rivers from Igor A. Shiklomanov, *World Water Resources: A New Appraisal and Assessment for the 21st Century* (Paris: International Hydrological Programme, UNESCO, 1998).

21. Jack E. Barbash, Research Chemist, USGS, e-mail to author, 16 November 1999.

22. UNEP, op. cit. note 1.

23. Sandra Postel, *Pillar of Sand* (New York: W.W. Norton & Company, 1999).

24. Sandra Postel, *Dividing the Waters: Food Security, Ecosystem Health, and the New Politics of Scarcity*, Worldwatch Paper 132 (Washington, DC: Worldwatch Institute, September 1996).

ICE COVER MELTING WORLDWIDE (pages 126–27)

1. Table 1 derived from the following sources: Arctic from Claire L. Parkinson et al., "Arctic Sea Ice Extents, Areas, and Trends, 1978–1996," *Journal of Geophysical Research*, 15 September 1999, from Ola M. Johannessen, Elena V. Shalina, and Martin W. Miles, "Satellite Evidence for an Arctic Sea Ice Cover in Transformation," *Science*, 3 December 1999, and from D.A. Rothrock, Y. Yu, and G.A. Maykut, "Thinning of the Arctic Sea-Ice Cover," *Geophysical Research Letters*, 1 December 1999; Greenland from William B. Krabill et al., "Rapid Thinning of Parts of the Southern Greenland Ice Sheet," *Science*, 5 March 1999; Columbia from "Alaska Glacier Travelling Quickly," *Environmental News Network*, 21 June 1999; Wilkins from National Snow and Ice Data Center (NSIDC), University of Colorado-Boulder, "Antarctic Ice Shelves Breaking Up Due to Decades of Higher Temperatures," press release (Boulder, CO: 7 April 1999); Tasman and Upsala from B. Blair Fitzharris et al., "The Cryosphere: Changes and Their Impacts," in Robert T. Watson et al. (eds.), *Climate Change 1995. Impacts, Adaptations and Mitigation of Climate Change: Scientific-Technical Analysis* (New York: Cambridge University Press, 1996); New Zealand, Caucasus, Tien Shan, Mt. Kenya, and Alps from "World's Glaciers Continue to Shrink," *Environmental News Network*, 27 May 1998; Gangotri from Mridula Chettri, "Glaciers Beating Retreat," *Down to Earth*, 30 April 1999, and from Fred Pearce, "Flooded Out," *New Scientist*, 5 June 1999; Glacier National Park from Alan Hall, "Going, Going—Gone?" *Scientific American*, 26 April 1999, and from Myrna H. P. Hall, *Predicting the Impact of Climate Change on Glacier and Vegetation Distribution in Glacier National Park to the Year 2100* (Syracuse: State University of New York: 1994), at U.S. Geological Survey, Glacier Field Station, <www.mesc.usgs.gov/glacier/glacier_model.html>, viewed 24 January 2000.

2. William K. Stevens, "1999 Continues Warming Trend Around Globe," *New York Times*, 19 December 1999.

3. Konstantin Y. Vinnikov et al., "Global Warming and Northern Hemisphere Sea Ice Extent," *Science*, 3 December 1999.

4. Bob Dickson, "All Change in the Arctic," *Nature*, 4 February 1999; Colin Woodard, "Glacial Ice is

Slip-Sliding Away," *Christian Science Monitor*, 10 December 1998.

5. United States from Curt Suplee, "Study: Arctic Sea Ice is Rapidly Dwindling," *Washington Post*, 3 December 1999; Ross from Parkinson et al., op. cit. note 1.

6. Rothrock, Yu, and Maykut, op. cit. note 1.

7. Figures from GLACIER, Rice University, "Introduction: How Big is the Ice?" <www.glacier.rice.edu/invitation/1_ice.html>, viewed 24 January 2000; disagreement from Malcolm W. Browne, "Under Antarctica, Clues to an Icecap's Fate," *New York Times*, 26 October 1999.

8. H. Conway et al., "Past and Future Grounding-Line Retreat of the West Antarctic Ice Sheet," *Science*, 8 October 1999.

9. Reed P. Scherer et al., "Pleistocene Collapse of the West Antarctic Ice Sheet," *Science*, 3 July 1998; ice streams from Ian Joughin et al., "Tributaries of West Antarctic Ice Streams Revealed by RADARSAT Interferometry," *Science*, 8 October 1999, and from Gabrielle Walker, "Great Rivers of Ice!" *New Scientist*, 17 April 1999.

10. NSIDC, op. cit. note 1; The Antarctica Project, "Ice Shelves," information sheet (Washington, DC: 24 June 1998).

11. The Antarctica Project, "Is West Antarctica Melting?" *The Antarctica Project Newsletter*, June 1998.

12. NSIDC, op. cit. note 1; Ted Scambos, NSIDC, discussion with author, 15 February 2000.

13. Curt Suplee, "Iceberg Bigger Than Delaware Breaks off Antarctica," *Washington Post*, 16 October 1999; "Renegade Iceberg Breaks Free From the Pack," *Environmental News Network*, 1 September 1999.

14. Fitzharris et al., op. cit. note 1.

15. World Glacier Monitoring Service, "Glacier Mass Balance Data 1997/98," *Mass Balance Bulletin No. 6* (Zurich: International Commission on Snow and Ice (ICSI), United Nations Environment Programme, and UNESCO, 1999.

16. Fitzharris et al., op. cit. note 1.

17. ICSI, cited in Chettri, op. cit. note 1.

18. Richard A. Kerr, "Will the Arctic Ocean Lose All Its Ice?" *Science*, 3 December 1999; cooling from Michael Kahn, "Shrinking Greenland Glacier Signals Global Warming," *Yahoo News*, 5 March 1999, and from Chettri, op. cit. note 1.

19, Arctic Monitoring and Assessment Programme, *Arctic Pollution Issues: A State of the Arctic Environment Report* (Oslo: 1997).

20. Carsten Rühlemann et al., "Warming of the Tropical Atlantic Ocean and Slowdown of Thermohaline Circulation During the Last Deglaciation," *Nature*, 2 December 1999; William K. Stevens, "Arctic Thawing May Jolt Sea's Climate Belt," *New York Times*, 7 December 1999.

21. Greenpeace International, *Climate Change and the Earth's Mountain Glaciers* (Amsterdam: May 1998).

22. Joby Warrick, "As Glaciers Melt, Talks on Warming Face Chill," *Washington Post*, 2 November 1998.

23. "Himalayan Glaciers Receding at Record Rates," *Global Environmental Change Report*, vol. 10, no. 18 (1998); effects and 500 million from Robert Marquand, "Glaciers in the Himalayas Melting at Rapid Rate," *Christian Science Monitor*, 5 November 1999.

24. "Melting Himalayan Glaciers Pose Flooding Dangers," *Reuters*, 3 June 1999.

25. Pearce, op. cit. note 1.

26. Sea levels from James E. Neumann et al., *Sea-level Rise & Global Climate Change*, prepared for Pew Center on Global Climate Change, February 2000; about half from Curt Suplee, "Icy Clues to Earth's Future," *Washington Post*, 21 February 1999.

27. Ian Allison, Roger Barry, and Barry Goodison, eds., *Science and Co-ordination Plan* (Tromsø, Norway: World Climate Research Programme's Climate and Cryosphere Project, 30 January 2000).

28. Mark Meier, Institute of Arctic and Alpine Research, University of Colorado at Boulder, discussion with author, 4 February 2000; Antarctica and Greenland from Greenpeace International, op. cit. note 21.

29. Figure of 70 percent from GLACIER, Rice University, "All About the Antarctic Ice Sheet," www.glacier.rice.edu/land/5_antarcticicesheetintro.html>, viewed 24 January 2000; size of ice mass from Woodard, op. cit. note 4; sea level rise estimates from GLACIER, Rice University, "More Than You Wished to Know About the GLACIER Web Site...," <www.glacier.rice.edu/misc/whatisglacier.html>, viewed 24 January 2000.

30. Peter N. Spotts, "Recording the Life Cycle of an Arctic Ice Sheet," *Christian Science Monitor*, 22 October 1998; Emily Laber, "Meltdown," *The Sciences*, July/August 1999.

31. "Sea Ice Retreat May Doom Arctic Wildlife," *Environmental News Network*, 27 October 1998.

STRESSES ON AMPHIBIANS GROW (pages 128–29)

1. Marcia Barinaga, "Where Have all the Froggies Gone?" *Science*, 2 March 1990.
2. Andrew R. Blaustein and David B. Wake, "The Puzzle of Declining Amphibian Populations," *Scientific American*, April 1995.
3. David B. Wake, "Declining Amphibian Populations," *Science*, 23 August 1991.
4. David B. Wake, "Action on Amphibians," *Trends in Ecology and Evolution*, 10 October 1998.
5. Table 1 from the following sources: Australia from William F. Laurance, Keith R. McDonald, and Richard Speare, "Epidemic Disease and the Catastrophic Decline of Australian Rain Forest Frogs," *Conservation Biology*, April 1996, and from Virginia Morell, "Are Pathogens Felling Frogs?" *Science*, 30 April 1999; Costa Rica from J. Alan Pounds et al., "Tests of Null Models for Amphibian Declines on a Tropical Mountain," *Conservation Biology*, December 1997, and from J. Alan Pounds, Michael P.L. Fogden, and John H. Campbell, "Biological Response to Climate Change on a Tropical Mountain," *Nature*, 15 April 1999; Charles A. Drost and Gary M. Fellers, "Collapse of a Regional Frog Fauna in the Yosemite Area of the California Sierra Nevada, USA," *Conservation Biology*, April 1996; Puerto Rico from Rafael L. Joglar and Patricia A. Burrowes, "Declining Amphibian Populations in Puerto Rico," in R. Powell and R.W. Henderson, eds., *Contributions to West Indian Herpetology: A Tribute to Albert Schwartz* (Ithaca, NY: The Society for the Study of Amphibians and Reptiles, 1996), and from Rafael L. Joglar, University of Puerto Rico, San Juan, e-mail to author, 1 March 2000.
6. Timothy R. Halliday and W. Ronald Heyer, "The Case of the Vanishing Frogs," *Technology Review*, May/June 1997.
7. Kathryn Phillips, "Prying into the Lives of Frogs," *National Wildlife*, October/November 1999.
8. James W. Petranka, Matthew E. Eldridge, and Katherine E. Haley, "Effects of Timber Harvesting on Southern Appalachian Salamanders," *Conservation Biology*, June 1993.
9. Cynthia Carey, "Hypothesis Concerning the Causes of the Disappearance of Boreal Toads from the Mountains of Colorado," *Conservation Biology*, June 1993; Laurance, McDonald, and Speare, op. cit. note 5.
10. Lee Berger et al., "Chytridiomycosis Causes Amphibian Mortality Associated with Population Declines in the Rain Forests of Australia and Central America," *Proceedings of the National Academy of Sciences*, 21 July 1998; Karen R. Lips, "Mass Mortality and Population Declines of Anurans at an Upland Site in Western Panama," *Conservation Biology*, February 1999; Morell, op. cit. note 5.
11. Ian Anderson, "Sick Pond Syndrome," *New Scientist*, 25 July 1998.
12. Pounds, Fogden, and Campbell, op. cit. note 5.
13. Christopher J. Still, Prudence N. Foster, and Stephen H. Schneider, "Simulating the Effects of Climate Change on Tropical Montane Cloud Forests," *Nature*, 15 April 1999.
14. Species introduction from G. M. Fellers and C. A. Drost, "Disappearance of the Cascades Frog *Rana cascadae*, at the Southern End of its Range, California, USA," *Biological Conservation*, vol. 65, no. 2 (1993), and from Peter B. Moyle, "Effects of Introduced Bullfrogs, *Rana catesbeiana*, on the Native Frogs of the San Joaquin Valley, California," *Copeia*, vol. 73, no. 1 (1973); UV-B radiation from Andrew Blaustein et al., "UV Repair and Resistance to Solar UV-B in Amphibian Eggs: A Link to Population Declines?" *Proceedings of the National Academy of Sciences*, March 1994, and from Andrew Blaustein et al., "Ambient UV-B Radiation Causes Deformities in Amphibian Embryos," *Proceedings of the National Academy of Sciences*, December 1997; acidification from T.J.C. Beebee et al., "Decline of the Natterjack Toad *Bufo calamita* in Britain: Palaeoecological, Documentary and Experimental Evidence for Breeding Site Acidification," *Biological Conservation*, vol. 53, no. 1 (1990), and from John Harte and Erika Hoffman, "Possible Effects of Acidic Deposition on a Rocky Mountain Population of the Tiger Salamander (*Ambystoma tigrinum*)" *Conservation Biology*, June 1989; agricultural pollution from Adolfo Marco, Consuelo Quilchano, and Andrew Blaustein, "Sensitivity to Nitrate and Nitrite in Pond-breeding Amphibians from the Pacific Northwest, USA," *Environmental Toxicology and Chemistry*, December 1999.
15. Tim Halliday, "A Declining Amphibian Conundrum," *Nature*, 30 July 1998.
16. Andrew R. Blaustein, et al., "Pathogenic Fungus Contributes to Amphibian Losses in the Pacific Northwest," *Biological Conservation*, vol. 67, no. 3 (1994).
17. Laurance, McDonald, and Speare, op. cit. note 5.

18. Vladimir Vershinin and Irina Kamkina, "Expansion of *Rana ridibunda* in the Urals—a Danger for Native Amphibians?" *Froglog* (newsletter of the Declining Amphibian Task Force), August 1999.
19. Robert C. Stebbins and Nathan W. Cohen, *A Natural History of Amphibians* (Princeton, NJ: Princeton University Press, 1995).
20. Laurie J. Vitt et al., "Amphibians as Harbingers of Decay," *BioScience*, June 1990.
21. James Hanken, "Why Are There So Many New Amphibian Species When Amphibians are Declining?" *Trends in Ecology and Evolution*, 1 January 1999.
22. Blaustein and Wake, op. cit. note 2.
23. Stebbins and Cohen, op. cit. note 19.
24. Richard Cannell, "Spawning a Painkiller," *Financial Times*, 17 April 1998; Andrew Pollack, "Biological Products Raise Genetic Ownership Issues," *New York Times*, 26 November 1999.
25. Tim Halliday, International Director, Declining Amphibian Task Force, World Conservation Union–IUCN, Open University, Milton Keynes, U.K., e-mail to author, 11 February 2000.

ENDOCRINE DISRUPTERS RAISE CONCERN (pages 130–31)

1. U.S. National Research Council (NRC), *Hormonally Active Agents in the Environment*, Committee on Hormonally Active Agents in the Environment, Board on Environmental Studies and Toxicology, Commission on Life Sciences (Washington, DC: National Academy of Sciences, prepublication copy July 1999).
2. Ted Schettler et al., *Generations at Risk: Reproductive Health and the Environment* (Cambridge, MA: The MIT Press, 1999).
3. U.S. Environmental Protection Agency, Office of Science Coordination and Policy, "Endocrine Disruptor Screening Program Web Site," <www.epa.gov/scipoly/oscpendo/index.htm>, viewed 20 December 1999.
4. World Resources Institute et al., *World Resources 1998–99* (New York: Oxford University Press, 1998).
5. René P. Schwarzenbach, Philip M. Gschwend, and Dieter M. Imboden, *Environmental Organic Chemistry* (New York: John Wiley & Sons, Inc., 1993).
6. Nigel Davis, Editor, *Chemical Insights*, e-mail to author, 13 January 2000.
7. Charles W. Schmidt, "Answering the Endocrine Test Questions," *Environmental Health Perspectives*, September 1999.
8. Theo Colborn, Frederick S. vom Saal, and Ana M. Soto, "Developmental Effects of Endocrine-Disrupting Chemicals in Wildlife and Humans," *Environmental Health Perspectives*, October 1993.
9. Frederick S. vom Saal and Daniel M. Sheehan, "Challenging Risk Assessment," *Forum for Applied Research and Public Policy*, fall 1998.
10. "Industry Glimpses New Challenges as Endocrine Science Advances," *ENDS Report*, March 1999.
11. Schmidt, op. cit. note 7; Christian G. Daughton and Thomas A. Ternes, "Pharmaceuticals and Personal Care Products in the Environment: Agents of Subtle Change?" *Environmental Health Perspectives*, December 1999; "Cosmetics and Food Preservatives Are Oestrogenic, Study Finds," *ENDS Report*, January 1999.
12. World Wildlife Fund, "Persistent Organic Pollutants: Hand-Me-Down Poisons that Threaten Wildlife and People," *Issue Brief* (Washington, DC: January 1999).
13. Susan Jobling, "A Variety of Environmentally Persistent Chemicals, Including Some Phthalate Plasticizers, Are Weakly Estrogenic," *Environmental Health Perspectives*, June 1995; "Toxic Toys," *Environmental Health Perspectives*, June 1999.
14. Theo Colborn, Dianne Dumonski, and John Peterson Myers, *Our Stolen Future* (New York: Penguin Group, 1996).
15. Weybridge Report, *European Workshop on the Impact of Endocrine Disrupters on Human Health and Wildlife, 2–4 December 1996, Weybridge, U.K.: Report of the Proceedings* (Copenhagen: European Commission DG XII, 16 April 1997).
16. R.M. Giusti, K. Iwamoto, and E.E. Hatch, "Diethylstilbestrol Revisited: A Review of the Long-term Health Effects," *Annals of Internal Medicine*, 15 May 1995; Schettler et al., op. cit. note 2.
17. Colborn, vom Saal, and Soto, op. cit. note 8.
18. Ibid.
19. Joseph L. Jacobson and Sandra W. Jacobson, "Intellectual Impairment in Children Exposed to Polychlorinated Biphenyls in Utero," *New England Journal of Medicine*, 12 September 1996.
20. Ibid.
21. S. Patandin et al., "Effects of Environmental Exposure to Polychlorinated Biphenyls and Dioxins on Cognitive Abilities in Dutch Children at 42 Months of Age," *Journal of Pediatrics*, January 1999.
22. Elisabeth Carlsen et al., "Evidence for

Decreasing Quality of Semen During Past 50 Years," *British Medical Journal*, 12 September 1992.

23. R. Bergstrom et al., "Increase in Testicular Cancer Incidence in Six European Countries: A Birth Cohort Phenomenon," *Journal of the National Cancer Institute*, June 1996; Leonard J. Paulozzi, "International Trends in Rates of Hypospadias and Cryptorchidism," *Environmental Health Perspectives*, April 1999.

24. NRC, op. cit. note 1; Shanna H. Swan, "Sperm Count Decline Studies Inconclusive," *Health and Environment Digest*, September 1998.

25. C.A. Paulsen et al., "Data from Men in the Greater Seattle Area Reveals No Downward Trend in Semen Quality: Further Evidence that Deterioration in Semen Quality Is Not Geographically Uniform," *Fertility and Sterility*, May 1996; H. Fisch et al., "Semen Analyses in 1,283 Men from the United States over a 25-Year Period: No Decline in Quality," *Fertility and Sterility*, May 1996; J. Ginsburg et al., "Residence in the London Area and Sperm Density," *Lancet*, 22 January 1994.

26. Schettler et al., op. cit. note 2.

PAPER RECYCLING REMAINS STRONG (pages 132–33)

1. U.N. Food and Agriculture Organization (FAO), *FAOSTAT Statistics Database*, < apps.fao.org >, viewed 30 October 1999).

2. Miller Freeman, Inc., *International Fact and Price Book 1999* (San Francisco: 1998).

3. Shushuai Zhu, David Tomberlin, and Joseph Buongiorno, *Global Forest Products Consumption, Production, Trade and Prices: Global Forest Products Model Projections to 2010* (Rome: Forest Policy and Planning Division, FAO, December 1998).

4. Miller Freeman, Inc., op. cit. note 2.

5. Paper Recycling Promotion Center, "Information about Paper Recycling Promotion Center" (Tokyo: n.d.).

6. Miller Freeman, Inc., op. cit. note 2.

7. Ibid.

8. "Divided EU Agrees on Packaging Directive, Joint Ratification of Climate Change Treaty," *International Environment Reporter*, 12 January 1994; "Council Agrees on FCP Phase-out, Packaging Recycling, Hazardous Waste List," *International Environment Reporter*, 11 January 1995; EC Directive from "Report Highlights Benefits of Increasing Newspaper Recycling,"

ENDS Report, July 1998.

9. Netherlands from "Paper, Cardboard Recycling Proceeding Well to Meet 2001 Targets, Industry Official Says," *International Environment Reporter*, 2 September 1998.

10. U.S. figure from Franklin Associates, Ltd., "Characterization of Municipal Solid Waste in the United States: 1998 Update," report prepared for the U.S. Environmental Protection Agency (EPA), Municipal and Industrial Solid Waste Division, Office of Solid Waste, Washington, DC, July 1999.

11. Ibid.

12. Discards of 44 million tons from Franklin Associates, Ltd., op. cit. note 10; amount of paper consumed in China (33 million tons) from Miller Freeman, Inc., op. cit. note 2.

13. Miller Freeman, Inc., op. cit. note 2.

14. Ibid.

15. Ibid.

16. Ibid.; 1940s from Maureen Smith, *The U.S. Paper Industry and Sustainable Production* (Cambridge, MA: The MIT Press, 1997).

17. FAO, op. cit. note 1.

18. Lauren Blum, Richard A. Denison, and John F. Ruston, "A Life-Cycle Approach to Purchasing and Using Environmentally Preferable Paper: A Summary of the Paper Task Force Report," *Journal of Industrial Ecology*, vol. 1, no. 3 (1997).

19. Ishiguro and Akiyama, 1994, cited in International Institute for Environment and Development (IIED), *Towards a Sustainable Paper Cycle* (London: 1996); Allen Hershkowitz, *Too Good To Throw Away: Recycling's Proven Record* (New York: Natural Resources Defense Council, February 1997).

20. Miller Freeman, Inc., op. cit. note 2.

21. Fred D. Iannazzi, "A Decade of Progress in U.S. Paper Recovery," *Resource Recycling*, June 1999; Susan Kinsella, "Recycled Paper Buyers, Where Are You?" *Resource Recycling*, November 1998.

22. U.K. study from "Report Highlights Benefits of Increasing Newspaper Recycling," *ENDS Report*, July 1998.

23. Ibid.

24. Process improvements from Jim Kenny, "More Recycled Fibers Generate Process Improvements," *Pulp and Paper Magazine*, June 1999, and from Marguerite Sykes et al., "Environmentally Sound Alternatives for Upgrading Mixed Office Wastes," *Proceedings of the 1995 International Environmental Conference*, Technical Association of the Pulp and Paper Industry, Atlanta, GA, 7–10 May 1995.

25. Figure of 43 percent from Miller Freeman, Inc., op. cit. note 2; potential recovery rate from Janet N. Abramovitz and Ashley T. Mattoon, *Paper Cuts: Recovering the Paper Landscape*, Worldwatch Paper 149 (Washington, DC: December 1999).

26. Abramovitz and Mattoon, op. cit. note 25.

ENVIRONMENTAL TREATIES GAIN GROUND (pages 134–35)

1. For the purposes of this tally, protocols and amendments to existing accords are considered to be new agreements. This listing includes regional and international binding accords. It does not include bilateral agreements or non-binding accords. This is in keeping with the criteria used in U.N. Environment Programme (UNEP), *Register of International Treaties and Other Agreements in the Field of the Environment 1996* (Nairobi: 1996), and in UNEP, *Register of International Treaties and Other Agreements in the Field of the Environment 1999* (Nairobi: forthcoming). A broader inventory that includes some bilateral and nonbinding accords lists nearly 900 international environmental instruments; see Edith Brown Weiss, Paul Szasz, and Daniel Magraw, *International Environmental Law: Basic Instruments and References* (Irvington-on-Hudson, NY: Transnational Juris Publications, Inc., 1992).

2. World Health Organization Europe, "WHO Ministerial Conference Takes Unprecedented Initiative on Water-related Disease, 35 European Countries Sign Protocol on Water and Health" press background (Copenhagen: 17 June 1999).

3. U.N. Economic Commission for Europe, "New Air Pollution Protocol to Save Lives and the Environment," press release (Geneva: 24 November 1999).

4. "The Protocol on Land-Based Sources of Marine Pollution in a Nutshell," <www.cep.unep.org/pubs/legislation/lbsmpnut.html>, viewed 1 March 2000.

5. For discussion of the rules governing the global commons, see Hilary French, *Vanishing Borders* (New York: W.W. Norton & Company, 2000).

6. "Montreal Protocol Gains Another Amendment," *Global Environmental Change Report*, 10 December 1999.

7. Secretariat of the Basel Convention, "Compensation and Liability Protocol Adopted by Basel Convention on Hazardous Wastes," press release (Geneva: 10 December 1999); "Agreement on Liability Protocol Reached at Basel Conference of Parties," *International Environment Reporter*, 8 December 1999.

8. Air pollution treaty from Marc Levy, "European Acid Rain: The Power of Tote-Board Diplomacy," *Institutions for the Earth* (Cambridge, MA: The MIT Press, 1993); chlorofluorocarbon production from Sharon Getamel, Dupont, Wilmington, DE, letter to Worldwatch, 15 February 1996, and from UNEP, *Data Report on Production and Consumption of Ozone Depleting Substances, 1986–1998* (Nairobi: 1999); The Antarctica Project, "The Protocol on Environmental Protection to the Antarctic Treaty," 14 June 1999, <www.asoc.org/currentpress/protocol.htm>, viewed 3 November 1999.

9. On the role of transparency, see Abram Chayes and Antonia Handler Chayes, *The New Sovereignty* (Cambridge, MA: Harvard University Press, 1995); Ronald B. Mitchell, "Sources of Transparency: Information Systems in International Regimes," *International Studies Quarterly*, vol. 42 (1998); and U.S. General Accounting Office (GAO), *International Environment: Literature on the Effectiveness of International Environmental Agreements* (Washington, DC: May 1999).

10. UNEP, "Establishment of Guidelines for the Second National Reports, Including Indicators and Incentive Measures," 15 December 1999, Fifth Meeting of the Subsidiary Body on Scientific, Technical and Technological Advice, Convention on Biological Diversity, Montreal, 31 January–4 February 2000.

11. UNEP, *Global Environmental Outlook 2000* (Nairobi: 1999).

12. GAO, op. cit. note 9; Rosemary Sandford, "International Environmental Treaty Secretariats; a Case of Neglected Potential?" *Environmental Impact Assessment Review*, vol. 198, no. 1 (1996); the 2000 budget for the Ramsar convention is only about $1.9 million, per "Financial and Budgetary Matters," Resolution No. VII.28 of 7th Meeting of the Conference of the Contracting Parties, Convention on Wetlands (Ramsar), San José, Costa Rica, 10–18 May 1999, while the 2000 budget for the climate convention is $11 million, per UNFCCC, "Administrative and Financial Matters: Programme Budget for the Biennium 2000–2001," Item 8 of the provisional agenda of the Fifth Session of the Conference of the Parties, Bonn, Germany, 25 October–5 November 1999;

the Clean Air program of the U.S. Environmental Protection Agency (EPA), for instance, had a budget of $541 million in fiscal year 2000, per EPA, *Summary of the 2001 Budget* (Washington, DC: January 2000).

13. See, for example, Duncan Brack, *International Trade and the Montreal Protocol* (London: Earthscan, for the Royal Institute of International Affairs, 1996).

14. Edith Brown Weiss and Harold K. Jacobson, *Engaging Countries: Strengthening Compliance with International Environmental Accords* (Cambridge, MA: The MIT Press, 1998).

ENVIRONMENTAL TAX SHIFTS MULTIPLYING (pages 138–39)

1. Table 1 is based on the following sources: Sweden description from P. Bohm, "Environment and Taxation: The Case of Sweden," in Organisation for Economic Co-operation and Development (OECD), *Environment and Taxation: The Cases of the Netherlands, Sweden and the United States* (Paris: 1994); Sweden quantity from Nordic Council of Ministers, *The Use of Economic Instruments in Nordic Environmental Policy* (Copenhagen: TemaNord, 1996); Denmark 1994 from Mikael Skou Andersen, "The Green Tax Reform in Denmark: Shifting the Focus of Tax Liability," *Journal of Environmental Liability*, vol. 2, no. 2 (1994); Spain description from Thomas Schröder, "Spain: Improve Competitiveness through an ETR," *Wuppertal Bulletin on Ecological Tax Reform* (Wuppertal, Germany: Wuppertal Institute for Climate, Environment, and Energy), summer 1995; Spain quantity from Juan-José Escobar, Ministry of Economy and Finance, Madrid, letter to author, 29 January 1997; Denmark 1996 from Ministry of Finance, *Energy Tax on Industry* (Copenhagen: 1995); Netherlands 1996 description from Ministry of Housing, Spatial Planning, and Environment (VROM), *The Netherlands' Regulatory Tax on Energy: Questions and Answers* (The Hague: 1996); Netherlands 1996 quantity from Koos van der Vaart, Ministry of Finance, The Hague, discussion with author, 18 December 1995; United Kingdom 1996 from "Landfill Tax Regime Takes Shape," *ENDS Report* (Environmental Data Services, London), November 1995; Finland from OECD, *Environmental Taxes and Green Tax Reform* (Paris: 1997); Germany, Italy, and France from

Stefan Speck and Paul Ekins, *Recent Developments in Environmental Taxation in EU Member States plus Norway and Switzerland*, draft report for European Commission (London: Forum for the Future, 2000); Netherlands 1999 from Arie Leder, Ministry of Finance, The Hague, e-mail to author, 17 February 2000; United Kingdom 2001 from Chris Hewett, Institute for Public Policy Research, London, discussion with author, 20 January 2000; total tax revenues for all countries from OECD, *Revenue Statistics of OECD Member Countries 1965–1997* (Paris: 1998).

2. Arthur Cecil Pigou, *The Economics of Welfare*, 4th ed. (London: Macmillan, 1932; first published 1920).

3. History from Ernst Ulrich von Weizsäcker, *Earth Politics* (London: Zed Books, 1994); see also Agnar Sandmo, "Optimal Taxation in the Presence of Externalities," *Swedish Journal of Economics*, vol. 77, no. 1 (1975), and Hans Christoph Binswanger et al., *Arbeit ohne Umweltzerstörung* (Frankfurt: Fischer, 1983), cited in von Weizsäcker, op. cit. this note.

4. Ernst U. von Weizsäcker and Jochen Jesinghaus, *Ecological Tax Reform* (London: Zed Books, 1992).

5. Thomas Sterner, *Environmental Tax Reform: The Swedish Experience* (Gothenburg, Sweden: Department of Economics, Gothenburg University, 1994); Svante Axelsson, Swedish Nature Protection Society, Stockholm, discussion with author, 29 January 1998.

6. Sterner, op. cit. note 5.

7. Bohm, op. cit. note 1; Nordic Council of Ministers, op. cit. note 1.

8. Swedish Environmental Protection Agency, *Environmental Taxes in Sweden: Economic Instruments of Environmental Policy* (Stockholm: 1997).

9. Andersen, op. cit. note 1.

10. Ministry of Finance, op. cit. note 1; VROM, op. cit. note 1; OECD, *Environmental Taxes and Green Tax Reform*, op. cit. note 1.

11. European Commission, Statistical Office of the European Communities (Eurostat), *European Economy*, (Luxembourg: Office for Official Publications of the European Communities), no. 64 (1997); idem, "Euro-zone Unemployment Stays at 9.8%: EU15 Unchanged at 9.0%," press release (Luxembourg: January 2000).

12. C.R. Bean, P.R.G. Layard, and S.J. Nickell, "The Rise in Unemployment: A Multi-Country Study," *Economica* 53, S1–S22, cited in OECD,

The OECD Jobs Study: Taxation, Employment and Unemployment (Paris: 1995).

13. Speck and Ekins, op. cit. note 1; Hewett, op. cit. note 1.

14. "Aufbruch und Erneuerung—Deutschlands Weg ins 21. Jahrhundert," Koalitionsvereinbarung zwischen der Sozialdemokratischen Partei Deutschlands und Bündnis90/Die Grünen, Bonn, Germany, 20 October 1998.

15. For a representative proposal using a broad-based energy tax, see Stefan Bach, Michael Kohlhaas, and Barbara Praetorius, "Ecological Tax Reform Even If Germany Has to Go It Alone," *Economic Bulletin* (Berlin: German Institute for Economic Research), July 1994; "German Ecological Tax Reform Plan Becomes Law," *Tax News Update* (Center for a Sustainable Economy, Washington, DC), 22 March 1999.

16. Kai Schlegelmilch, Ministry of Environment, Berlin, e-mail to author, 1 September 1999.

17. David Malin Roodman, *The Natural Wealth of Nations* (New York: W.W. Norton & Company, 1998).

18. Ibid.

SATELLITES BOOST ENVIRONMENTAL KNOWLEDGE (pages 140–41)

1. Christopher A. Legg and Yves Laumonier, "Fires in Indonesia, 1997: A Remote Sensing Perspective," *Ambio*, September 1999; Nigel Dudley, *The Year the World Caught Fire* (Gland, Switzerland: World Wide Fund for Nature, December 1997).

2. Margot Cohen and Murray Hiebert, "Where's There's Smoke... Spread of Indonesian Oil-Palm Plantations Fuels the Haze," *Far Eastern Economic Review*, 2 October 1997.

3. Claire L. Parkinson, *Earth From Above: Using Color-Coded Satellite Images to Examine the Global Environment* (Sausalito, CA: University Science Books, 1997).

4. National Geographic, *National Geographic Satellite Atlas of the World* (Washington, DC: National Geographic Society Book Division, 1998).

5. Draft Report of the Third United Nations Conference on the Exploration and Peaceful Uses of Outer Space, 16 April 1999.

6. Thomas M. Lillesand and Ralph W. Kiefer, *Remote Sensing and Image Interpretation*, 3rd ed. (New York: John Wiley and Sons, 1994).

7. Charles Kennel, Pierre Morel, and Gregory Williams, "Keeping Watch on the Earth: An Integrated Observing Strategy," *Consequences*, vol. 3, no. 2 (1997).

8. Parkinson, op. cit. note 3.

9. R.B. Myneni et al., "Increase in Plant Growth in the Northern High Latitudes from 1981–1991," *Nature*, 17 April 1997; Michael Oppenheimer, "Global Warming and the Stability of the West Antarctic Ice Sheet," *Nature*, 28 May 1998.

10. Dana Mackenzie, "Ocean Floor Is Laid Bare by New Satellite Data," and Walter H.F. Smith and David T. Sandwell, "Global Sea Floor Topography from Satellite Altimetry and Ship Depth Soundings," *Science*, 26 September 1997.

11. David Herring, *The First EOS Satellite: NASA's Earth Observing System, EOS AM-1* (Greenbelt, MD: NASA Goddard Space Flight Center, 1998).

12. Salley Bowen, "Peru Introduces the Science of Fishing—By Satellite," *Financial Times*, 27 August 1999.

13. "Italians to Spot Toxic Waste from Space," *Associated Press*, 25 October 1999.

14. "Sea WiFS Completes a Year of Remarkable Earth Observations," *Science Daily*, 18 September 1998.

15. Vernon Loeb, "Spy Satellite Will Take Photos for Public Sale," *Washington Post*, 25 September 1999. Government spy satellites produce images of comparable or sharper detail.

16. Andrew Backover, "Civilian Firm in Orbit: Space Imaging Wins 3-Way Race," *Denver Post*, 25 September 1999.

17. Bjørn Willum, "From Kosovo to Chechnya, Selling Images from Above," *Christian Science Monitor*, 30 November 1999.

18. Ann Florini, "The End of Secrecy," *Foreign Policy*, summer 1998; see also Ann M. Florini and Yahya Dehqnzada, "Commercial Satellite Imagery Comes of Age," *Issues in Science and Technology*, fall 1999.

CORPORATE MERGERS SKYROCKET (pages 142–43)

1. Value of mergers based on data received from Carrie Smith, Thomson Financial Securities Data, Newark, NJ, e-mail to author, 28 January 2000. These and all other merger data are concerned with announced, not actually consummated, deals. Since some intended takeovers do not go ahead because they are called off by the companies involved or challenged by governmental oversight bodies, there may be a slight difference between announced and actually

completed mergers. This discrepancy, however, is too small to invalidate the trends and order of magnitude of the data discussed here.

2. Smith, op. cit.note 1.

3. Ibid.

4. Ibid.

5. Ibid.

6. Worldwatch calculation, based on ibid.

7. U.N. Conference on Trade and Development (UNCTAD), *World Investment Report 1999* (New York: 1999).

8. "Faites Vos Jeux," *The Economist*, 4 December 1999.

9. Growth of strategic partnerships discussed in UNCTAD, op. cit. note 7.

10. Joshua Karliner, *The Corporate Planet* (San Francisco, CA: Sierra Club Books, 1997); Jerry Mander and Edward Goldsmith, eds., *The Case Against the Global Economy* (San Francisco, CA: Sierra Club Books, 1996); William Greider, *One World, Ready Or Not* (New York: Touchstone Books, 1997).

11. Steve Lohr, "Behemoths in a Jack-Be-Nimble Economy," *New York Times*, 12 September 1999.

12. Laura M. Holson and Seth Schiesel, "MCI to Buy Sprint in Swap of Stock for $108 Billion," *New York Times*, 5 October 1999. The takeover proposal was subsequently valued at $130 billion.

13. Saul Hansell, "America Online Agrees to Buy Time Warner for $165 Billion; Media Deal is Richest Merger," *New York Times*, 11 January 2000; Edmund L. Andrews and Andrew Ross Sorkin, "$183 Billion Deal in Europe to Join 2 Wireless Giants," *New York Times*, 4 February 2000.

14. Thomson Financial Securities Data, "YTD Market Totals," < www.secdata.com >, viewed 16 March 2000.

15. Hansell, op. cit. note 13; Lawrie Mifflin, "Viacom to Buy CBS, Forming 2d Largest Media Company," *New York Times*, 8 September 1999; Andrew Ross Sorkin and Melody Petersen, "Glaxo and SmithKline Agree to Form Largest Drugmaker," *New York Times*, 18 January 2000; Melody Petersen, "Pfizer Gets its Deal to Buy Warner-Lambert for $90.2 Billion," *New York Times*, 8 February 2000; Andrews and Sorkin, op. cit. note 13; Allen R. Myerson, "Exxon and Mobil Announce $80 Billion Deal to Create World's Largest Company," *New York Times*, 2 December 1998; Keith Bradsher, "Capacity Glut Likely to Spur More Auto Mergers," *New York Times*, 14 November 1998; Edmund L. Andrews, "The Gossip of Europe: Talk About Mergers Surrounds Automakers," *New York Times*, 17 February 2000; Claudia H. Deutsch, "Another Big Paper Company Acquisition," *New York Times*, 23 February 2000.

16. Smith, op. cit. note 1.

17. Ibid.

18. Thomson Financial Securities Data, "The World is Not Enough," press release, < www.sec data.com/news/news_corp/archive/2000/press0 1_05_00.html >, 5 January 2000.

19. "The Path to the MCI Worldcom and Sprint Merger," *New York Times*, 7 October 1999. The merger has been challenged by the U.S. Federal Trade Commission and by the European Commission's antitrust office; Edmund L. Andrews, "European Regulators Frown On a Combined MCI-Sprint," *New York Times*, 22 February 2000.

20. "The Net Gets Real," *The Economist*, 15 January 2000; "European Media: Flirtation and Frustration," *The Economist*, 11 December 1999; Janine Jaquet, "The Media Nation: TV," *The Nation*, 8 June 1998; "The New Global Media" (special issue), *The Nation*, 29 November 1999.

21. Robert W. McChesney, *Rich Media, Poor Democracy* (Chicago: University of Illinois Press, 1999).

22. UNCTAD, op. cit. note 7.

23. Thomson Financial Securities Data, op. cit. note 18.

24. "Europe's New Capitalism: Bidding for the Future," *The Economist*, 12 February 2000.

25. UNCTAD, op. cit. note 7; Louis Uchitelle, "As Mergers Multiply, So Does the Danger," *New York Times*, 13 February 2000.

26. UNCTAD, op. cit. note 7; Uchitelle, op. cit. note 25.

27. McChesney, op. cit. note 21.

28. Not all cross-border mergers are financed by foreign direct investment (FDI), and data are lacking to establish a clear relationship between crossborder mergers and FDI worldwide. In 1998, 90 percent of investments by foreign investors in the United States went to finance mergers and acquisitions. In the same year, the value of cross-border mergers in South, East, and Southeast Asia accounted for 16 percent of total FDI inflows. UNCTAD, op. cit. note 7.

29. UNCTAD, op. cit. note 7.

30. Ibid.

31. Ibid.

32. Affiliates' sales from ibid.; world exports from "World Trade Stable in Value" in this volume.

33. UNCTAD, op. cit. note 7.

WIND ENERGY JOBS RISING
(pages 146–47)

1. For example, the executive council of the U.S. labor union umbrella organization AFL-CIO issued a statement opposing the Kyoto Protocol on climate change, arguing that it "could have a devastating impact on the U.S. economy and American workers"; AFL-CIO Executive Council, "U.S. Energy Policy" (Washington, DC: 17 February 1999).
2. Bundesverband Wind Energie e.V., *Windenergie—25 Fakten* (Osnabrück, Germany: 1999).
3. Ibid.
4. Worldwatch projection, based on various sources cited here.
5. The average size of wind turbines installed grew roughly three- to fourfold during the 1990s, and is expected to double again during the next decade to about 1.5 megawatts per plant, and even larger sizes for the emerging offshore wind industry. European Wind Energy Association (EWEA), Forum for Energy and Development, and Greenpeace International, *Wind Force 10: A Blueprint to Achieve 10% of the World's Electricity from Wind Power by 2020* (London: 1999).
6. Danish Wind Turbine Manufacturers Association, "Employment in the Wind Power Industry," *Wind Power Notes No. 2*, March 1996.
7. EWEA estimate from European Commission, "Energy for the Future: Renewable Sources of Energy," White Paper for a Community Strategy and Action Plan, COM(97)599 final (26/11/97).
8. EWEA, Forum for Energy and Development, and Greenpeace, op. cit. note 5.
9. Danish Wind Turbine Manufacturers Association, op. cit. note 6.
10. EWEA, Forum for Energy and Development, and Greenpeace, op. cit. note 5.
11. Ibid.
12. Figure of 86,000 jobs is a Worldwatch estimate, based on a 1999 installation figure of 3,900 megawatts, which is a Worldwatch estimate (see "Wind Power Booms" in this volume), and 22 jobs per megawatt.
13. EWEA, Forum for Energy and Development, and Greenpeace, op. cit. note 5.
14. European Commission, Directorate-General for Energy, "Wind Energy—The Facts," Vol. 3 (Brussels, 1997).
15. Greenpeace Germany, "Solar-Jobs 2010: Neue Arbeitsplätze durch neue Energien," summary of an April 1997 study, <www.greenpeace.de/GP_DOK_30/STU_KURZ>, viewed 2 August 1999.
16. "The Wind Energy Industry—Status and Prospects," EWEA, <www.ewea.org/industry.htm>, viewed 17 February 2000.
17. In 1999, an estimated 29 terawatt-hours were produced; EWEA, Forum for Energy and Development, and Greenpeace, op. cit. note 5.
18. European firms' market share from U.S. General Accounting Office, *Renewable Energy: DOE's Funding and Markets for Wind Energy and Solar Cell Technologies* (Washington, DC: May 1999).
19. India turbine manufacturers from Christopher Flavin, "Wind Power Blows to New Record," in Lester R. Brown, Michael Renner, and Brian Halweil, *Vital Signs 1999* (New York: W.W. Norton & Company, 1999); spare parts and maintenance from Raman Thothathri, "The Wind Brought Jobs and Prosperity," *New Energy*, November 1999.
20. Peter Korneffel, "The Lull Before the Storm," *New Energy*, May 1999.
21. EWEA, Forum for Energy and Development, and Greenpeace, op. cit. note 5.
22. Ibid.
23. Ibid.

TUBERCULOSIS RESURGING WORLDWIDE (pages 148–49)

1. World Health Organization (WHO), *World Health Report 1999* (Geneva: 1999).
2. WHO, "Tuberculosis Fact Sheet," <www.who.int/gtb/publications/factsheet/index.htm>, viewed 16 December 1999.
3. U.S. Centers for Disease Control and Prevention (CDC), "Questions and Answers About TB 1994," <www.cdc.gov/nchstp/tb/faqs/qa.htm>, viewed 14 January 2000.
4. Ibid.
5. Ibid.
6. WHO, op. cit. note 2.
7. WHO, "TB is Single Biggest Killer of Young Women," press release (Geneva: May 1998); WHO, op. cit. note 1.
8. WHO, op. cit. note 7.
9. WHO, op. cit. note 1.
10. "Medical Notes: Tuberculosis," *BBC News Online*, 26 May 1998; WHO, op. cit. note 1.
11. WHO, op. cit. note 1.
12. "Medical Notes: Tuberculosis," op. cit. note 10; CDC, *Reported Tuberculosis in the United States,*

1998 (Atlanta, GA: 1999).

13. WHO, "Sharp Rise in Tuberculosis Strikes Eastern Europe," press release (Geneva: March 1998).

14. WHO, op. cit. note 1.

15. Ibid.; WHO, *TB: A Crossroads. WHO Report on the Global Tuberculosis Epidemic 1998* (Geneva: 1998).

16. WHO, op. cit. note 2.

17. WHO, op. cit. note 1.

18. Ibid.

19. WHO, op. cit. note 15; WHO, op. cit. note 2.

20. WHO, *Anti-Tuberculosis Drug Resistance in the World* (Geneva: 1997).

21. WHO, op. cit. note 2; WHO, op. cit. note 20.

22. WHO, op. cit. note 15.

23. WHO, *Tuberculosis and Air Travel: Guidelines for Prevention and Control* (Geneva: 1998).

24. John Donnelly and Dave Montgomery, "Experts Warn of Global TB Epidemic if Threat Not Attacked Now," *Miami Herald*, 17 March 1999.

25. WHO, op. cit. note 15.

26. WHO, op. cit. note 2.

27. Figure of 10 million from Dermot Maher et al., *Guidelines for the Control of Tuberculosis in Prisons* (Geneva: WHO, 1998): 30 million from Ian Smith, "TB is an Issue of Human Rights," *TB Treatment Observer* (WHO), 15 May 1999.

28. Alex Goldfarb, "'Gulag' Strains Pose New Epidemic Threat," *TB Treatment Observer* (WHO), 15 May 1999.

29. WHO, *Guidelines for the Control of Tuberculosis in Prisons* (Geneva: 1998).

30. Ibid.

31. Tuong Nguyen, "Tuberculosis: A Global Health Emergency," *Outlook* (Program for Appropriate Technology in Health), November 1999.

32. WHO, "Tuberculosis," <www.who.int/gpv-dvacc/diseases/TB.htm>, viewed 16 December 1999.

33. Nguyen, op. cit. note 31; "Novel Approach Brings TB Success," *BBC News Online*, 14 July 1999.

34. WHO, op. cit. note 2.

35. Nguyen, op. cit. note 31; WHO, op. cit. note 2.

36. WHO, op. cit. note 1.

37. WHO, op. cit. note 15; WHO, *Global Tuberculosis Control 1999* (Geneva: 1999).

38. WHO, "DOTS: Directly Observed Treatment Short-course," <www.who.int/gtb/dots/index.htm>, viewed 16 December 1999; WHO, op. cit. note 15.

39. WHO, op. cit. note 15.

40. WHO, op. cite. note 1; WHO, op. cit. note 15.

41. WHO, op. cit. note 13; WHO, op. cit. note 15.

42. Nguyen, op. cit. note 31; Judith Miller, "Study Says New TB Strains Need an Intensive Strategy," *New York Times*, 28 October 1999.

PRISON POPULATIONS EXPLODING (pages 150–51)

1. Roy Walmsley, "World Population Prison List," Research Findings No. 88 (London: Development and Statistics Directorate, Home Office Research, 1999).

2. Figure of 10 million from Dermot Maher et al., *Guidelines for the Control of Tuberculosis in Prisons* (Geneva: World Health Organization (WHO), 1998); 30 million from Ian Smith, "TB is an Issue of Human Rights," *The TB Treatment Observer* (WHO), 15 May 1998.

3. Human Rights Watch, "Excessive Pretrial Detention," <hrw.org/advocacy/prisons/pretrial.htm>, viewed 16 February 2000; Charles Thomas, professor of sociology (retired), University of Florida, <web.crim.ufl.edu/pcp/census/1999/Market.html>, viewed 16 February 2000.

4. Walmsley, op. cit. note 1.

5. Ibid.

6. Ibid.

7. Worldwatch calculation based on ibid.

8. Walmsley, op. cit. note 1.

9. Vivien Stern, *A Sin Against the Future: Imprisonment in the World* (Boston: Northeastern University Press, 1998).

10. Moscow Center for Prison Reform, <www.mcpr.org>, viewed 22 February 2000.

11. Stern, op. cit. note 9.

12. Ibid.

13. Ibid.

14. Ibid.

15. Jenni Gainsborough, The Sentencing Project, Washington, DC, e-mail to author, 28 February 2000.

16. The Sentencing Project, "Facts about Prisons and Prisoners," <www.sentencingproject.org>, viewed 24 February 2000.

17. Marc Mauer, "Americans Behind Bars: U.S. and International Use of Incarceration, 1995," at Families Against Mandatory Minimums, <www.famm.org/victims8.htm>, viewed 24 February 2000.

18. Eric Schlosser, "The Prison-Industrial Complex," *The Atlantic Monthly*, December 1998.

19. Cases from Smith, op. cit. note 2; Russia from Stern, op. cit. note 9.

20. Laura M. Maruschak, *HIV in Prisons 1997* (Washington, DC: Bureau of Justice Statistics, November 1999).

21. Thomas, op. cit. note 3.

22. Ibid.

23. Ibid.

24. Schlosser, op. cit. note 18.

25. Wages from ibid.; health care from Gainoborough, op. cit. note 15.

26. Schlosser, op. cit. note 18.

27. "Arizona Shows the Way on Drugs," *New York Times*, 24 April 1999.

28. Ibid.

29. Brendan O'Friel, "Crime and Punishment," *Resurgence*, November/December 1998.

30. "Remarks of the Honorable Janet Reno, Attorney General of the United States on Re-entry Court Initiative," address at the John Jay College of Criminal Justice in New York, 10 February 2000.

31. Schlosser, op. cit. note 18.

32. "Remarks of the Honorable Janet Reno," op. cit. note 30.

WOMEN SLOWLY GAIN GROUND IN POLITICS (pages 152–53)

1. Inter-Parliamentary Union (IPU), "Women in National Parliaments: Situation as of 5 December 1999," <www.ipu.org>, viewed 22 December 1999.

2. IPU, *Women in Parliaments 1945–1995: A World Statistical Survey* (Geneva: 1995).

3. Jodi L. Jacobson, *Gender Bias: Roadblock to Sustainable Development*, Worldwatch Paper 110 (Washington, DC: Worldwatch Institute, September 1992).

4. IPU, op. cit. note 2.

5. "Women's Day in Kuwait," *The Economist*, 22 May 1999; Douglas Jehl, "Debate on Women's Rights Shows Deep Rift in Kuwait Society," *New York Times*, 20 December 1999.

6. IPU, op. cit. note 2.

7. Robert Marquand, "Women at Pinnacle of Power," *Christian Science Monitor*, 1 May 1998.

8. Agence France Presse, "Latvia Woman Is President," *New York Times*, 18 June 1999.

9. Seth Mydans, "Can She Run Indonesia? It's About Islam, or Is It?" *New York Times*, 18 June 1999.

10. Seth Mydans, "Indonesia Opposition Leader Chosen as the Vice President," *New York Times*, 22 October 1999.

11. "Busting Turkey's Grey Male Monotony," *The Economist*, 17 April 1999.

12. Andrea Mandel-Campbell, "Mexican Machismo Takes Mauling from Women: A String of Female Political Successes Have Rocked the Male Establishment," *Financial Times*, 12 October 1999.

13. Stryker McGuire with Barry White, "Breaking into the Club," *Newsweek* (Atlantic edition), 29 June 1998.

14. Jeff Israely, "In Italy, Women's Political Star Rises," *Boston Globe*, 5 June 1999.

15. South African cabinet cited on WomenWatch, The UN Internet Gateway on the Advancement and Empowerment of Women, <www.un.org/womenwatch/index.html>, viewed 2 January 2000.

16. Linda Feldman, "Election Coverage of Women: More on Personality, Less on Issues," *Christian Science Monitor*, 25 October 1999.

17. Seema Nayar, "Hidden Behind the Saris," *Newsweek*, 16 February 1998.

18. Jack Epstein, "'Lipstick Lobby' Puts 75,000 Women on the Ballot in Patriarchal Brazil," *Christian Science Monitor*, 10 September 1996.

19. Gail Russell Chaddock, "Quotas Boost Women Pols," *Christian Science Monitor*, 14 May 1997.

20. Christina Nifong, "Women Link Up from Austria to Zambia," *Christian Science Monitor*, 19 April 1995.

21. Sisterhood is Global, <www.sigi.org>, viewed 2 January 2000.

22. Barbara Crossette, "Women's Rights Gaining Attention Within Islam," *New York Times*, 12 May 1996.

23. Marilyn Gardner, "World Leaders' Council: Only Women Need Apply," *Christian Science Monitor*, 8 October 1997; John F. Kennedy School of Government, "Kennedy School Honors the Establishment of the Council of Women World Leaders," press release, undated, <www.ksg.harvard.edu/ksgpress/ksg_news/announcemts/council.htm>.

24. Women 2000, "Beijing + Five at a Glance," <www.un.org/womenwatch/followup/beijing5/index.html>, viewed 7 February 2000.

25. Platform for Action on the Internet, <www.un.org/womenwatch/daw/beijing/platform/>, viewed 7 February 2000.

THE VITAL SIGNS SERIES

Some topics are included each year in Vital Signs; *others, particularly those in Part Two, are included only in certain years. The following is a list of the topics covered thus far in the series, with the year or years each appeared indicated in parentheses.*

Part One: KEY INDICATORS

FOOD TRENDS
  Grain Production (1992–2000)
  Soybean Harvest (1992–2000)
  Meat Production (1992–2000)
  Fish Catch (1992–2000)
  Grain Stocks (1992–99)
  Grain Used for Feed (1993, 1995–96)
  Aquaculture (1994, 1996, 1998)

AGRICULTURAL RESOURCE
TRENDS
  Grain Area (1992–93, 1996–97, 1999–2000)
  Fertilizer Use (1992–2000)
  Irrigation (1992, 1994, 1996–99)
  Grain Yield (1994–95, 1998)
  Pesticide Trade (2000)

ENERGY TRENDS
  Oil Production (1992–96, 1998)
  Wind Power (1992–2000)
  Nuclear Power (1992–2000)
  Solar Cell Production (1992–2000)
  Natural Gas (1992, 1994–96, 1998)

Energy Efficiency (1992)
Geothermal Power (1993, 1997)
Coal Use (1993–96, 1998)
Hydroelectric Power (1993, 1998)
Carbon Use (1993)
Compact Fluorescent Lamps (1993–96,
  1998–2000)
Fossil Fuel Use (1997, 1999–2000)

ATMOSPHERIC TRENDS
  CFC Production (1992–96, 1998)
  Global Temperature (1992–2000)
  Carbon Emissions (1992, 1994–2000)

ECONOMIC TRENDS
  Global Economy (1992–2000)
  Third World Debt (1992–95, 1999–2000)
  International Trade (1993–96, 1998–2000)
  Steel Production (1993, 1996)
  Paper Production (1993, 1994, 1998–2000)
  Advertising Expenditures (1993, 1999)
  Roundwood Production (1994, 1997, 1999)
  Gold Production (1994, 2000)

Television Use (1995)
Storm Damages (1997–2000)
U.N. Finances (1998–99)
Tourism (2000)

TRANSPORTATION TRENDS
Bicycle Production (1992–2000)
Automobile Production (1992–2000)
Air Travel (1993, 1999)
Motorbike Production (1998)

ENVIRONMENTAL TRENDS
Pesticide Resistance (1994)
Sulfur and Nitrogen Emissions (1994–97)
Environmental Treaties (1995)
Nuclear Waste (1995)

COMMUNICATION TRENDS
Satellite Launches (1998–99)
Telephones (1998–2000)
Internet Use (1998–2000)

SOCIAL TRENDS
Population Growth (1992–2000)
Cigarette Production (1992–2000)
Infant Mortality (1992)
Child Mortality (1993)
Refugees (1993–2000)
HIV/AIDS Incidence (1994–2000)
Immunizations (1994)
Urbanization (1995–96, 1998, 2000)
Life Expectancy (1999)
Polio (1999)

MILITARY TRENDS
Military Expenditures (1992, 1998)
Nuclear Arsenal (1992, 1994–96, 1999)
Arms Trade (1994)
Peace Expenditures (1994–2000)
Wars (1995, 1998–2000)
Armed Forces (1997)

Part Two: SPECIAL FEATURES

ENVIRONMENTAL FEATURES
Bird Populations (1992, 1994)
Forest Loss (1992, 1994–98)
Soil Erosion (1992, 1995)
Steel Recycling (1992, 1995)
Nuclear Waste (1992)
Water Scarcity (1993)
Forest Damage from Air Pollution (1993)
Marine Mammal Populations (1993)
Paper Recycling (1994, 1998, 2000)
Coral Reefs (1994)
Energy Productivity (1994)
Amphibian Populations (1995, 2000)
Large Dams (1995)
Water Tables (1995, 2000)
Lead in Gasoline (1995)

Aquatic Species (1996)
Environmental Treaties (1996, 2000)
Ecosystem Conversion (1997)
Primate Populations (1997)
Ozone Layer (1997)
Subsidies for Environmental Harm (1997)
Tree Plantations (1998)
Vertebrate Loss (1998)
Organic Waste Reuse (1998)
Nitrogen Fixation (1998)
Acid Rain (1998)
Transgenic Crops (1999–2000)
Pesticide Resistance (1999)
Algal Blooms (1999)
Urban Air Pollution (1999)
Biomass Energy (1999)

Organic Agriculture (2000)
Groundwater Quality (2000)
Ice Melting (2000)
Endocrine Distrupters (2000)

AGRICULTURAL FEATURES
Pesticide Control (1996)
Organic Farming (1996)

ECONOMIC FEATURES
Wheat/Oil Exchange Rate (1992, 1993)
Trade in Arms and Grain (1992)
Cigarette Taxes (1993, 1995, 1998)
U.S. Seafood Prices (1993)
Environmental Taxes (1996, 1998, 2000)
Private Finance in Third World (1996, 1998)
Storm Damages (1996)
Aid for Sustainable Development (1997)
Food Aid (1997)
R&D Expenditures (1997)
Urban Agriculture (1997)
Electric Cars (1997)
Arms Production (1997)
Fossil Fuel Subsidies (1998)
Metals Exploration (1998)
Pollution Control Markets (1998)
Urban Transportation (1999)
Transnational Corporations (1999–2000)
Government Corruption (1999)
Satellite Monitoring (2000)

SOCIAL FEATURES
Income Distribution (1992, 1995, 1997)
Maternal Mortality (1992, 1997)
Access to Family Planning (1992)
Literacy (1993)
Fertility Rates (1993)
Traffic Accidents (1994)

Life Expectancy (1994)
Women in Politics (1995, 2000)
Computer Production and Use (1995)
Breast and Prostate Cancer (1995)
Homelessness (1995)
Hunger (1995)
Access to Safe Water (1995)
Infectious Diseases (1996)
Landmines (1996)
Violence Against Women (1996)
Voter Turnouts (1996)
Aging Populations (1997)
Noncommunicable Diseases (1997)
Extinction of Languages (1997)
Female Education (1998)
Sanitation (1998)
Unemployment (1999)
Nongovernmental Organizations (1999)
Malnutrition (1999)
Sperm Count (1999)
Fast-Food Use (1999)
Jobs in Wind Energy (2000)
Tuberculosis (2000)
Prison Populations (2000)

MILITARY FEATURES
Nuclear Arsenal (1993)
U.N. Peacekeeping (1993)
Small Arms (1998–99)